HIGH SPEED SEMICONDUCTOR DEVICES

HIGH SPEED SEMICONDUCTOR DEVICES

Circuit aspects and fundamental behaviour

H. Beneking

Former Head of The Institute for Semiconductor Electronics at the Technical University of Aachen, Germany

CHAPMAN & HALL

London · Glasgow · Weinheim · New York · Tokyo · Melbourne · Madras

Published by Chapman & Hall, 2–6 Boundary Row, London SE1 8HN, UK

Chapman & Hall, 2–6 Boundary Row, London SE1 8HN, UK

Blackie Academic & Professional, Wester Cleddens Road, Bishopbriggs, Glasgow G64 2NZ, UK

Chapman & Hall GmbH, Pappelallee 3, 69469 Weinheim, Germany

Chapman & Hall USA, One Penn Plaza, 41st Floor, New York NY 10119, USA

Chapman & Hall Japan, ITP-Japan, Kyowa Building, 3F, 2-2-1 Hirakawacho, Chiyoda-ku, Tokyo 102, Japan

Chapman & Hall Australia, Thomas Nelson Australia, 102 Dodds Street, South Melbourne, Victoria 3205, Australia

Chapman & Hall India, R. Seshadri, 32 Second Main Road, CIT East, Madras 600 035, India

First edition 1994

Typeset in 10/12 pts Times by Thomson Press (India) Ltd., New Delhi
Printed in Great Britain by St. Edmundsbury Press Ltd., Bury St. Edmunds, Suffolk

ISBN 0 412 56220 0

A catalogue record for this book is available from the British Library

Library of Congress Catalog Card Number: 94–70972

∞ Printed on permanent acid-free text paper, manufactured in accordance with ANSI/NISO Z39.48-1992 and ANSI/NISO Z39.48-1984

Contents

Preface

This book is devoted to the electronics of advanced high speed semiconductor devices. It gives a comprehensive introduction to the circuit theory involved for solid state physicists as well as for electrical engineers. To take advantage of the content the reader should be familiar with the general behaviour and the circuit application of electronic devices.

Semiconductor devices are the basis of electronic systems used in engineering and related fields. The designer has to know their practical needs as well as their physical limits. The demands from an application point of view have to be correlated with the technological feasibility including reliability considerations, reproducibility and low cost fabrication. New device structures have to be checked regarding their usefulness under these auspices.

Here an overview of the general electrotechnical properties of semiconductor devices is given. Important electrical data are introduced, relevant to circuit design and electrical device characterization.

The appendices and also the exercises contain additional information and detailed calculations not included in the main text. For this purpose, and to become familiar with the content of the book, exercises are placed at the end of each chapter, while the solutions of all exercises can be found at the end of the book. A serious treatment of them will allow the reader to gain practical experience.

The book will fulfil its purpose if it gives the reader easier access and better evaluation of future trends and enables him/her to become creative in designing and applying new devices, not in a sense of unrealistic developments but on the basis of physical-technological limitations and practical needs.

For critical reading of the manuscript I am indebted to my colleague Hans-Jürgen Schmitt, and for assistance in preparing the exercises Volker Sommer, Antonio Mesquida Küsters, Bernhard Opitz and further support of the Institute for Semiconductor Electronics of the Technical University Aachen. The preparation of the figures was performed by Rüdiger Tuzinski, and finally I have to thank Sigrun Lessmann for typing the manuscript.

H. Beneking
Aachen, 1993

Introduction

The devices that we shall consider have to be applicable to practical signal processing either for analog or digital signals. In the first case their frequency response, linearity, narrowband and broadband behaviour are of importance, in the latter case their time domain behaviour, low jitter and bit-rate dependence. For digital communication at a bit rate R the uppermost transmitted frequency f_{max} has to be $f_{max} > 3R$, to achieve pulse transmission instead of sinusoidal signals, and because of the signal distribution between 0 and 1 the low frequency limit f_{min} of the digital transmission has to be at least $f_{min} < R/100$, if not DC.

On the other hand, for narrowband analog signals at a frequency $f = f_0$ only the device behaviour in the narrow frequency band $B = 2\Delta f$ with $f_0 \pm \Delta f, \Delta f \ll f_0$, is of importance.

Table 1 summarizes some of the viewpoints regarding the two groups of applications. They are dependent on the system configurations, e.g. signal processing, satellite communication, short distance computer interconnects as local area networks (LANs), long haul optoelectronic transmission, radio astronomy, etc. In Fig. 1 the frequency bands of interest are indicated, including their designation for communication purposes. Modern active devices are capable of covering at least the K_u band; however, above 60 GHz FET oscillators and amplifiers and THz oscillations in quantum well structures have been reported as well as switching times below 10 ps at room temperature, down to about 2 ps.

Regarding the bit rates to be considered one has to recall that with 7 bit for one character corresponding to one ASCII sequence (or 8 bit = 1 byte with parity) a data transfer of 1 Gbit s^{-1} is capable of transmitting about 60 000 pages of text per second – one typed page contains about 2500 characters requiring 17 500 bits – or up to about 50 TV channels, depending on the type of modulation.

Electro-optical signal processing becomes more and more important: experimental links using monomode optical fibres up to about 4 Gbit s^{-1} have already been established, whereas device structures in the 10 Gbit s^{-1} range are being tested in laboratories. The heterodyne concept and the implementation of (erbium-doped) fibre amplifiers are capable of extending the repeater distances as well as the number of channels which can be transmitted, leading to a new era of optoelectronic communication.

All conceivable systems are composed of building blocks containing electronic devices. Therefore the starting point of any development is the circuit element

Table 1 Considerations for analog and digital device applications

General considerations

- Temperature range
  - Ge, $Ga_{0.47}In_{0.53}As$ — $\vartheta_{max} \approx 80°C$
  - Si — $\vartheta_{max} \approx 120°C$
  - GaAs, InP — $\vartheta_{max} \approx 300°C$
- Reliability
- Uniformity
- Chip size
- Integration
  hybrid or monolithic
- Costs

Analog applications	**Digital applications**
• Frequency range	• Low or high data range
Silicon $f < 4$–$6\,GHz$	Silicon $R < 1\,Gbit\,s^{-1}$
III–V materials $f > 3$–$6\,GHz$	III–V materials $R > 1\,Gbit\,s^{-1}$
• Narrow or wide frequency band	• Frontends
• Frontends	sensitivity
noise equivalent power	bit error rate
noise matching	• Digital circuits
cross-modulation	jitter
• Generators	reflections
amplitude noise	delay time
phase noise	power delay product
• Amplifiers	pulse shape
bandwidth	turn-on time
linearity	turn-off time
output power	storage time
1′ dB compression point	• Soft error immunity
power-added efficiency	
power consumption	
• Failure rate	

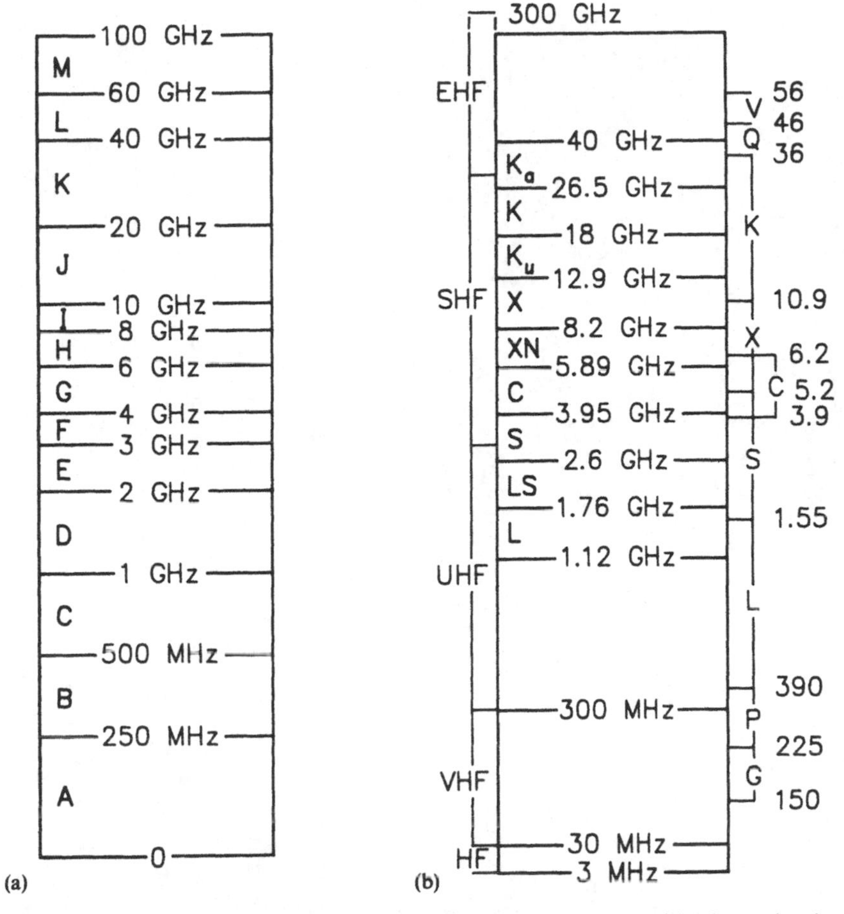

Fig. 1 Frequency bands: (a) new band allocations; (b) waveguide denominations (left) and band allocations (right).

and its electrical characterization as discussed in the following chapters. Photonic devices are not covered separately, because the emphasis is on electronic devices. Their electric behaviour is equivalent, as explained in section 5.6. The additional photonic noise is shown in section F.6 (appendices).

1

Electrical parameters and equivalent circuits

The active devices have to be integrated into an electrical circuit forming a link in an amplifier chain where the signal processing is performed. The systems can be generalized corresponding to Fig. 1.1. The electrical part is positioned between the sensor and the actuator which link the electronics to the outside world, e.g. pressure input and sound output, or optical signal input and optical signal output in the case of optoelectronic repeaters.

For the mathematical description and analysis of the active devices and passive components, application-related parameters have been developed. They allow not only a mathematical representation of the devices themselves but also the analysis of the whole circuit where the devices are implemented. Computer-aided design (CAD) tools are available, and the application of advanced circuit theory allows us to deal with complex and nonlinear functional dependencies in the frequency and time domain (e.g. Fourier analysis, Fourier transformation).

An introduction to large signal effects and their analysis is given in Chapter 6. Here the fourpole and wave parameters are presented as applicable to the small signal case. Furthermore the corresponding equivalent circuits and the graphical representation of the resulting dependencies are shown.

1.1 FOURPOLES

If sinusoidal signals are considered which include in principle all possible waveforms via Fourier series or Fourier transformation, the commonly used linearized version of the electric part is shown in Fig. 1.2. The 'active device' can be interpreted as the whole amplifier or one part of it, the 'generator' and 'load' as the output of the foregoing and input of the following stage, respectively. In discussing the electronic behaviour of an electronic device generally, one single stage incorporating this (active) device is considered.

The arrows in Fig. 1.2 indicate the positive directions for currents and voltages, e.g. $V_1 = +2\,\mathrm{V}$ corresponds to 'point A is $+2\,\mathrm{V}$ more positive than point B', or $I_1 = -1\,\mathrm{mA}$ corresponds to a current flow of 1 mA across point A out of the fourpole, which essentially flows back across point B. These directions have to be defined by the arrows; however, they can be chosen arbitrarily.

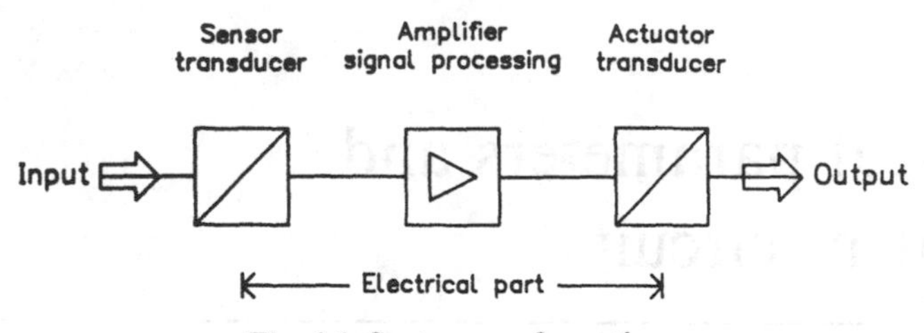

Fig. 1.1 System configuration.

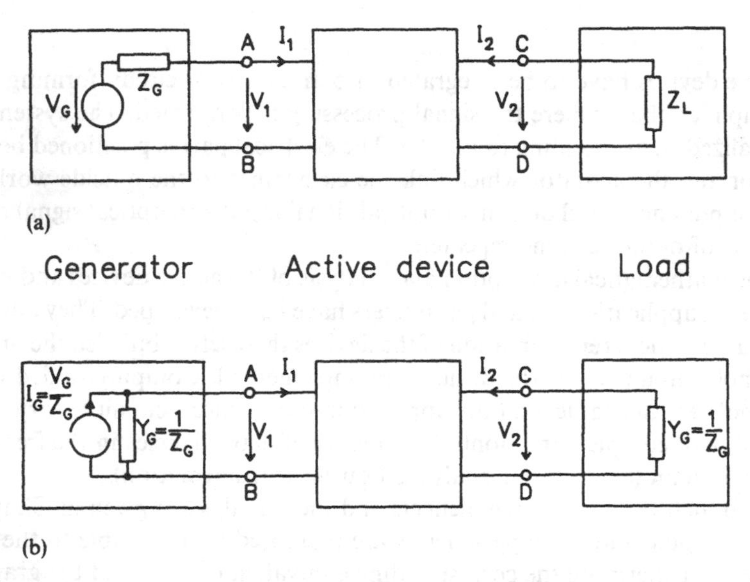

Fig. 1.2 Electrical part of a system: active device between generator and load. (a) Impedance notation of generator and load equivalent circuit; (b) admittance notation of generator and load equivalent circuit.

As indicated in Fig. 1.2 the generator and load are represented by equivalent circuits, a linearized version of the electrical behaviour of both components. The same can be done for the active device, where the equivalent circuit has to be derived from physical-technological considerations and its correspondence to the fourpole equations. Because four quantities I_1, V_1, I_2, V_2 have to be considered, different fourpole equations can be formed. They are shown in Appendix A, where their application for coupled fourpoles is also demonstrated, and where their transformations are listed.

In the low frequency limit the fourpole parameters correspond to the partial differential quotients of the internal DC dependencies of the device under consideration. To derive these quasistatic parameters the static characteristics have to be expressed by differential series. This will be shown here in case of a field-effect transistor where, e.g. the DC currents $I_G = I_G(V_{GS}, V_{DS})$ and $I_D = I_D(V_{GS}, V_{DS})$ have to be expressed by twodimensional differential Taylor series around the bias point $\{V_{GSP}, V_{DSP}\}$ with the corresponding currents I_{GP} and I_{DP}, respectively. In this case the currents are expressed as functions of the voltages, leading to the admittance parameters y_{kl} of the device. If other independent variables are used, other combinations follow, e.g. with $V_1 = V_1(I_1, V_2)$ and $I_2 = I_2(I_1, V_2)$ the hybrid parameters h_{kl}, etc. (Appendix A).

With $I_G = I_G(V_{GS}, V_{DS})$ it follows that

$$\begin{aligned} I_G = I_{GP} &+ \left(\frac{\partial I_G}{\partial V_{GS}}\right)(V_{GS} - V_{GSP}) + \left(\frac{\partial I_G}{\partial V_{DS}}\right)(V_{DS} - V_{DSP}) \\ &+ \frac{1}{2}\left(\frac{\partial^2 I_G}{\partial V_{GS}^2}\right)(V_{GS} - V_{GSP})^2 + \frac{1}{2}\left(\frac{\partial^2 I_G}{\partial V_{DS}^2}\right)(V_{DS} - V_{DSP})^2 \\ &+ \left(\frac{\partial I_G}{\partial V_{GS}}\right)\left(\frac{\partial I_G}{\partial V_{DS}}\right)(V_{GS} - V_{GSP})(V_{DS} - V_{DSP}) + \cdots \end{aligned} \tag{1.1}$$

and with $I_D = I_D(V_{GS}, V_{DS})$

$$\begin{aligned} I_D = I_{DP} &+ \left(\frac{\partial I_D}{\partial V_{GS}}\right)(V_{GS} - V_{GSP}) + \left(\frac{\partial I_D}{\partial V_{DS}}\right)(V_{DS} - V_{DSP}) \\ &+ \frac{1}{2}\left(\frac{\partial^2 I_D}{\partial V_{GS}^2}\right)(V_{GS} - V_{GSP})^2 + \frac{1}{2}\left(\frac{\partial^2 I_D}{\partial V_{DS}^2}\right)(V_{DS} - V_{DSP})^2 \\ &+ \left(\frac{\partial I_D}{\partial V_{GS}}\right)\left(\frac{\partial I_D}{\partial V_{DS}}\right)(V_{GS} - V_{GSP})(V_{DS} - V_{DSP}) + \cdots \end{aligned} \tag{1.2}$$

Setting

$$\left.\begin{aligned} i_g &= I_G - I_{GP} \\ i_d &= I_D - I_{DP} \\ v_{gs} &= V_{GS} - V_{GSP} \\ v_{ds} &= V_{DS} - V_{DSP} \end{aligned}\right\} \tag{1.3}$$

the deviations from the bias condition can be interpreted as the applied quasistatic signals under the given bias conditions.

By neglecting higher order terms (small signal case) it follows that

$$\left.\begin{aligned} i_g &= \left(\frac{\partial I_G}{\partial V_{GS}}\right)v_{gs} + \left(\frac{\partial I_G}{\partial V_{DS}}\right)v_{ds} \\ i_d &= \left(\frac{\partial I_D}{\partial V_{GS}}\right)v_{gs} + \left(\frac{\partial I_D}{\partial V_{DS}}\right)v_{ds} \end{aligned}\right\} \tag{1.4}$$

As can be seen, the partial derivatives represent the corresponding quasistatic admittance parameters of the fourpole. These are the input conductance (output short-circuited)

$$\left.\frac{\partial I_{\mathrm{G}}}{\partial V_{\mathrm{GS}}}\right|_{V_{\mathrm{DS}}=\mathrm{const.}\,\hat{=}\,v_{\mathrm{ds}}=0} = g_{11} = g_{\mathrm{i}} \tag{1.5}$$

the reverse mutual transconductance (mutual conductance, input short-circuited)

$$\left.\frac{\partial I_{\mathrm{G}}}{\partial V_{\mathrm{DS}}}\right|_{V_{\mathrm{GS}}=\mathrm{const.}\,\hat{=}\,v_{\mathrm{gs}}=0} = g_{12} = g_{\mathrm{r}} \tag{1.6}$$

forward transconductance (mutual conductance, output short-circuited)

$$\left.\frac{\partial I_{\mathrm{D}}}{\partial V_{\mathrm{GS}}}\right|_{V_{\mathrm{DS}}=\mathrm{const.}\,\hat{=}\,v_{\mathrm{ds}}=0} = g_{21} = g_{\mathrm{m}} \tag{1.7}$$

and the output conductance (input short-circuited)

$$\left.\frac{\partial I_{\mathrm{D}}}{\partial V_{\mathrm{DS}}}\right|_{V_{\mathrm{GS}}=\mathrm{const.}\,\hat{=}\,v_{\mathrm{gs}}=0} = g_{22} = g_{\mathrm{o}}. \tag{1.8}$$

These equations can be extended to higher frequencies, if complex notations are used,

$$g_{\mathrm{kl}} \rightarrow y_{\mathrm{kl}} = g_{\mathrm{kl}} + jb_{\mathrm{kl}} \tag{1.9}$$

and introducing complex currents I and voltages, V, respectively, with the notation

$$\left.\begin{aligned} I &= \hat{I}\,\mathrm{e}^{\mathrm{j}\varphi_{\mathrm{I}}} \\ V &= \hat{V}\,\mathrm{e}^{\mathrm{j}\varphi_{\mathrm{V}}} \end{aligned}\right\} \tag{1.10}$$

where $\mathrm{j} = (-1)^{1/2}$. These are phasors, leading to the real time-dependent currents and voltages at the given frequency $f = \omega/2\pi$ by the real parts of the corresponding time-dependent phasors,

$$\left.\begin{aligned} i(\mathrm{t}) &= \mathrm{Re}\{I\,\mathrm{e}^{\mathrm{j}\omega t}\} \\ v(\mathrm{t}) &= \mathrm{Re}\{V\mathrm{e}^{\mathrm{j}\omega t}\} \end{aligned}\right\} \tag{1.11}$$

The quantities $\hat{I}$, $\hat{V}$ represent the amplitudes.

If DC values have to be considered, capital letters are used as subscripts, e.g. V_{P} for the DC bias voltage at the 'working point' P; contrary to this V_{cd} refers to an alternating voltage between the connection points C, D, etc.

With this notation the admittance matrix equation becomes

$$\begin{bmatrix} I_1 \\ I_2 \end{bmatrix} = \begin{bmatrix} y_{11} & y_{12} \\ y_{21} & y_{22} \end{bmatrix} \begin{bmatrix} V_1 \\ V_2 \end{bmatrix} \tag{1.12}$$

or in symbolic notation

$$\boldsymbol{I} = \boldsymbol{y} \cdot \boldsymbol{V} \tag{1.13}$$

which has to be read as

$$\left.\begin{aligned} I_1 &= y_{11} V_1 + y_{12} V_2 \\ I_2 &= y_{21} V_1 + y_{22} V_2 \end{aligned}\right\} \tag{1.14}$$

The parameters y are the complex admittance parameters of the fourpole, $y_{kl} = g_{kl} + jb_{kl}$, where g_{kl} is the real part and b_{kl} the imaginary part. They correspond to a complex admittance $Y = G + jB$, where G is the conductance and B the susceptance. Analogously a complex impedance $Z = R + jX$ consists of a real part (resistance R) and an imaginary part (reactance X), where the inverse, $1/Z = Y = G + jB$, is the equivalent admittance (Fig. 1.2).

At moderately high frequencies (around 0.01–1 GHz), generally the y-parameters are used, if the RF behaviour of an active device is to be characterized. From a measurement point of view they are very well suited because the device has to be short-circuited if the y-parameters are measured, corresponding to a stable condition without self-oscillations (section 4.1.2 and Appendix D). However, in the upper GHz range transmission lines are used rather than conventional wiring. In this case voltages and currents can also be defined but not measured in a conventional way. Better suited parameters are the scattering parameters S_{kl} and the transmission parameters T_{kl}, in which the common fourpole parameters can be converted (Appendix A). They characterize in an equivalent manner the fourpole which then has to be interpreted as a twoport with incoming and outgoing waves (Fig. 1.3). The geometrical position of the planes 1 and 2 of the twoports are of importance for the values of the wave parameters, at least regarding their phase.

This wave concept (Carlin, 1956) which permits us to measure relatively simply the corresponding parameters at very high frequencies, will be discussed next. Here no multiport devices are considered; the wave concept is applied only to twoports. It should be mentioned that multiports can be treated in a similar manner, by using multipole (or multiport) parameters. Then the corresponding matrices contain more elements, e.g. nine in case of a sixpole (threeport) configuration (e.g. the representation of a mixer stage with two inputs (signal and local oscillator) and one output (intermediate frequency)).

In practice, in the upper GHz range waveguides, striplines and coplanar transmission lines or slotlines and finline configurations are used. Coaxial cables

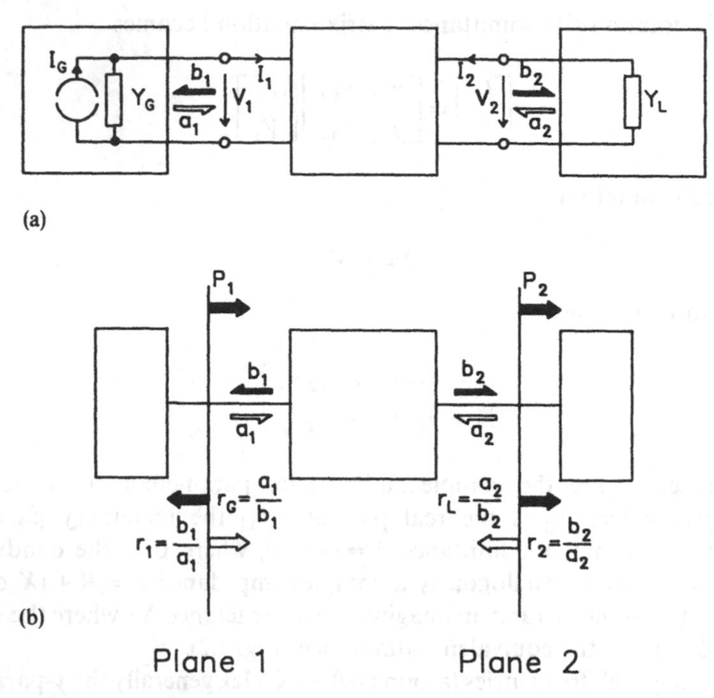

Fig. 1.3 A fourpole as twoport. (a) Characterization with incoming (a_1, a_2 and outgoing (b_1, b_2) waves; (b) wave notation including generator (source) and load.

which would allow conventional description by voltages are limited to frequencies below 40 GHz because of parasitic modes related to the inner diameter of the cables. A survey with many references can be found in Hoffmann (1987), whereas collections of related publications on microwave integration (monolithic microwave integrated circuits, MMICs) have been published by Artech House (Frey, 1975; Frey and Bhasin, 1985) and IEEE Press (Pucel, 1985).

1.2 THE WAVE CONCEPT

Using transmission lines the transported power can be considered by discussing the power flow in forward and reverse direction. Instead of the complex voltages and currents along the line, waves a_v, b_v are defined which correspond to these power flows (Fig. 1.3).

The (complex) voltage V at a given geometrical plane can be expressed by the superposition of both the wave voltages V^+, V^-, corresponding to the power flows, by

$$V = V^+ + V^- \tag{1.15}$$

and the current I at the given position in positive direction by

$$I = I^+ - I^- \tag{1.16}$$

With the characteristic transmission line impedance Z_0 this is equivalent to

$$V = Z_0(I^+ + I^-) \tag{1.17}$$

and

$$I = \frac{1}{Z_0}(V^+ - V^-) \tag{1.18}$$

respectively.

The power transported in the positive direction, P^+ is

$$P^+ = \frac{|V^+|^2}{2Z_0} \tag{1.19}$$

The power transported in the negative direction, P^- is

$$P^- = \frac{|V^-|^2}{2Z_0} \tag{1.20}$$

Therefore the waves are defined as the complex quantities

$$a = \frac{V^+}{(Z_0)^{1/2}} \tag{1.21}$$

and

$$b = \frac{V^-}{(Z_0)^{1/2}} \tag{1.22}$$

which is equivalent to

$$a = I^+(Z_0)^{1/2} \tag{1.23}$$

and

$$b = I^-(Z_0)^{1/2} \tag{1.24}$$

respectively.

The apparent voltages V and currents I in the transmission line can be introduced into the equations for a and b, which allows us to verify the power flow equation

$$P = P^{+} - P^{-} \tag{1.25}$$

Because

$$\left.\begin{aligned} V &= V^{+} + V^{-} = a(Z_0)^{1/2} + b(Z_0)^{1/2} \\ I &= I^{+} - I^{-} = \frac{a}{(Z_0)^{1/2}} - \frac{b}{(Z_0)^{1/2}} \end{aligned}\right\} \tag{1.26}$$

it follows that

$$\left.\begin{aligned} \frac{V}{(Z_0)^{1/2}} &= a + \mathrm{b} \\ I(Z_0)^{1/2} &= a - b \end{aligned}\right\} \tag{1.27}$$

or

$$a = \frac{1}{2}\left\{\frac{V}{(Z_0)^{1/2}} + I(Z_0)^{1/2}\right\} \tag{1.28}$$

and

$$b = \frac{1}{2}\left\{\frac{V}{(Z_0)^{1/2}} - I(Z_0)^{1/2}\right\} \tag{1.29}$$

Introducing these expressions in the power flow equation (1.25),

$$P = \frac{aa^*}{2} - \frac{bb^*}{2} \tag{1.30}$$

($A^* = |A|^2/A$ conjugated complex value of A) it follows that

$$\begin{aligned} P &= \frac{1}{8}\left\{\frac{|V|^2}{Z_0} + |I|^2 Z_0 + V^*I + VI^* - \frac{|V|^2}{Z_0} - |I|^2 Z_0 + V^*I + VI^*\right\} \\ &= \tfrac{1}{4}\{V^*I + VI^*\} \\ &= \tfrac{1}{2}\mathrm{Re}\{VI^*\} \end{aligned} \tag{1.31}$$

This is the expression for the transmitted power.

With the definition of the reflection coefficient

$$r = \frac{V^{-}}{V^{+}} \tag{1.32}$$

which is equivalent to

$$r = \frac{b}{a} \tag{1.33}$$

the power flow can also be expressed by

$$P = \frac{aa^*}{2}\left\{1 - \frac{bb^*}{aa^*}\right\} = P^+(1 - |r|^2) \tag{1.34}$$

If no reflection occurs, $b = 0$, $P^- = 0$ and therefore $r = 0$, the maximum power transmission becomes possible (matching condition). Therefore the available power is

$$P_{AV} = P^+ \tag{1.35}$$

(also Appendix B).

1.2.1 Scattering parameters

In Fig. 1.3 the embedded (active) device is shown in two versions, corresponding to the two different but equivalent descriptions. Figure 1.3(a) shows the conventional fourpole configuration, Fig. 1.3(b) the twoport configuration, applicable to the wave concept.

In the latter case, instead of the y-parameters (or the other V, I related parameters), wave parameters have to be used. These are the

- scattering parameters S_{kl}
- transmission parameters T_{kl}

Again four independent parameters, which can be arranged in a matrix configuration as in the case of the conventional fourpole parameters, characterize the device completely.

The scattering matrix $\boldsymbol{S}$ is defined by the matrix equation

$$\boldsymbol{b} = \boldsymbol{S} \cdot \boldsymbol{a} \tag{1.36}$$

or

$$\left.\begin{aligned} b_1 &= S_{11}a_1 + S_{12}a_2 \\ b_2 &= S_{21}a_1 + S_{22}a_2 \end{aligned}\right\} \tag{1.37}$$

and the transmission matrix $\boldsymbol{T}$ by

$$\left.\begin{aligned} b_1 &= T_{11}a_2 + T_{12}b_2 \\ a_1 &= T_{21}a_2 + T_{22}b_2 \end{aligned}\right\} \tag{1.38}$$

The latter is of importance for cascaded twoports (Appendix A). The meaning of the S_{kl} parameters can simply be derived from the definition given by the above equations (Fig. 1.3(b)).

It follows that

$$S_{11} = \left.\frac{b_1}{a_1}\right|_{a_2=0} = \text{input reflection } r_1|_{r_L=0} \tag{1.39}$$

if the output is matched,

$$S_{12} = \left.\frac{b_1}{a_2}\right|_{a_1=0} = \text{reverse transmission} \tag{1.40}$$

defined/measured at the matched input port,

$$S_{21} = \left.\frac{b_2}{a_1}\right|_{a_2=0} = \text{forward transmission} \tag{1.41}$$

for matched output,

$$S_{22} = \left.\frac{b_2}{a_2}\right|_{a_1=0} = \text{output reflection } r_2|_{r_G=0} \tag{1.42}$$

if the input is matched.

The input and output reflections for arbitrary load/generator follow by inspecting the twoport relationships to

$$r_1 = \frac{b_1}{a_1} = S_{11} + S_{12}\frac{a_2}{a_1} \tag{1.43}$$

and with

$$a_2 = b_2 r_L \tag{1.44}$$

to

$$\frac{1}{r_L} = \frac{b_2}{a_2} = S_{21}\frac{a_1}{a_2} + S_{22} \tag{1.45}$$

Therefore

$$\frac{a_2}{a_1} = S_{21}\frac{1}{(1/r_L) - S_{22}} \tag{1.46}$$

which leads to

$$r_1 = S_{11} + \frac{S_{12}S_{21}}{(1/r_L) - S_{22}} \tag{1.47}$$

and analogously

$$r_2 = S_{22} + \frac{S_{12}S_{21}}{(1/r_G) - S_{11}} \tag{1.48}$$

As in the case of conventional fourpoles, where the parameters with the index 12 characterize the feedback (Appendix A), here S_{12} is responsible for the dependence of $r_1(r_L)$ and $r_2(r_G)$, respectively.

Because the squares $|a_v|^2$, $|b_v|^2$ correspond to the transported powers, the quotients of these squares are power relationships of the twoport.

The quantity

$$|S_{11}|^2 = \left.\frac{P_1^-}{P_1^+}\right|_{r_L=0} = |r_1|^2|_{r_L=0} \tag{1.49}$$

gives the ratio of the reflected to incident power at the input for matched output (input power reflection).

The forward transmission S_{21} leads to

$$|S_{21}|^2 = \left.\left|\frac{b_2}{a_1}\right|^2\right|_{a_2=0} = \left.\frac{P_L}{P_1^+}\right|_{a_2=0 \,\hat{=}\, r_L=0} \tag{1.50}$$

This characterizes the forward power transfer of the twoport, measurable for matched load ($r_L = 0$) (section 2.1).

Correspondingly

$$|S_{12}|^2 = \left.\left|\frac{b_1}{a_2}\right|^2\right|_{a_1=0} \tag{1.51}$$

is the reverse power transfer, and

$$|S_{22}|^2 = \left.\left|\frac{b_2}{a_2}\right|^2\right|_{a_1=0 \,\hat{=}\, r_G=0} \tag{1.52}$$

is the output power reflection for matched input.

These power ratios can be expressed via the decibel notation

$$\frac{P_\alpha/P_\beta}{\text{dB}} = 10 \log\left\{\frac{P_\alpha}{P_\beta}\right\} \tag{1.53}$$

leading to

$$\frac{|S_{kl}|^2}{\text{dB}} = 10\log\{S_{kl}|^2\} \tag{1.54}$$

To apply the decibel notation generally to the S-parameters is incorrect because only the squares $|S_{kl}|^2$ are power-related quantities (section 1.3.2).

1.3 GRAPHICAL REPRESENTATION

There are several possibilities to show graphically the complex fourpole parameters, the twoport parameters or simply the frequency behaviour of a given admittance and impedance, respectively. Those of interest for the RF characterization of a device will be introduced in this chapter.

1.3.1 Fourpole parameters

The fourpole parameters, e.g. y_{kl}, h_{kl}, z_{kl}, are complex quantities with or without dimensions, being a ratio or corresponding to an admittance or impedance (Appendix A). Because their values are complex and frequency dependent, they can be represented graphically using a two-dimensional plane, as common for conventional complex numbers. Three versions are in use which can also be used

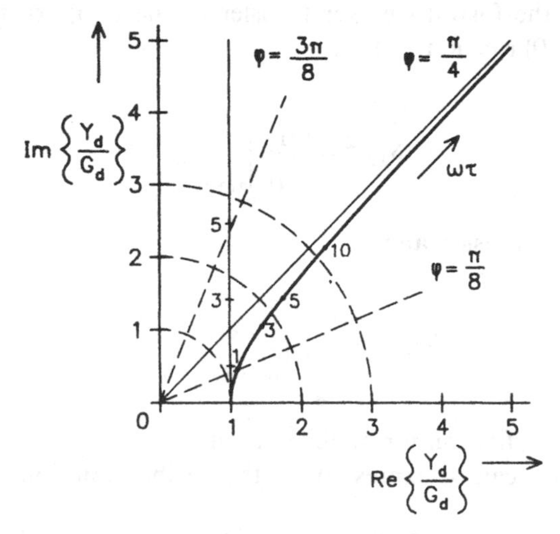

Fig. 1.4 Admittance plane $\{B, G\}$ with graphical representation of a p^+–n diode. Rectangular and polar coordinates show real and imaginary parts and the magnitude/phase angle, respectively.

for the graphical representation of fourpole parameters and twoport parameters, respectively. They will first be demonstrated for a simple p–n diode.

As an example of the first version the small signal input admittance $Y = G + jB$ of a long-base p^+–n diode is shown in the B–G plane (Fig. 1.4) (base width $W \gg$ diffusion length L_p; p-type doping $N_A \gg$ n-type doping N_D).

The ideal device, neglecting parasitics and the Schottky capacitance C_S in parallel to the junction, shows an admittance function given by

$$Y_d = G_d(1 + j\omega\tau)^{1/2} \tag{1.55}$$

where

$$G_d = \frac{I_P}{V_T}$$

is the diffusion conductance with I_P forward direct current (bias), $V_T = kT/q$ thermal voltage, and τ minority carrier lifetime (e.g. Feldmann, 1972) ($V_T \approx 26\,\text{mV}$ at room temperature, τ in the ns to ms range).

With

$$Y_d = \begin{cases} G_d + j\omega\frac{\tau}{2}G_d & \text{for} \quad \omega\tau \ll 1 \\ G_d\left(\frac{\omega\tau}{2}\right)^{1/2} + jG_d\left(\frac{\omega\tau}{2}\right)^{1/2} & \text{for} \quad \omega\tau \gg 1 \end{cases} \tag{1.56}$$

and

$$\begin{aligned} |Y_d| &= G_d(1 + \omega^2\tau^2)^{1/4} \\ &= \begin{cases} G_d\left\{1 + \left(\frac{\omega\tau}{2}\right)^2\right\} & \text{for} \quad \omega\tau \ll 1 \\ G_d(\omega\tau)^{1/2} & \text{for} \quad \omega\tau \gg 1 \end{cases} \end{aligned} \tag{1.57}$$

$$\begin{aligned} \varphi_{Y_d} &= \arctan\left\{\frac{\text{Im}(1 + j\omega\tau)^{1/2}}{\text{Re}(1 + j\omega\tau)^{1/2}}\right\} \\ &= \begin{cases} \arctan\left(\frac{\omega\tau}{2}\right) & \text{for} \quad \omega\tau \ll 1 \\ \arctan(1) = \frac{\pi}{4} & \text{for} \quad \omega\tau \gg 1 \end{cases} \end{aligned} \tag{1.58}$$

the graphical construction and its representation in Fig. 1.4 can easily be understood. As usual a normalization to $Y_d|_{\omega=0} = G_d$ is made, to show the general behaviour. As can be seen from the indicated parameter values ($\omega\tau$), the low frequency version of the admittance curve is valid up to $\omega\tau \approx 1$, whereas the 45° curve has to be applied for $\omega\tau > 10$.

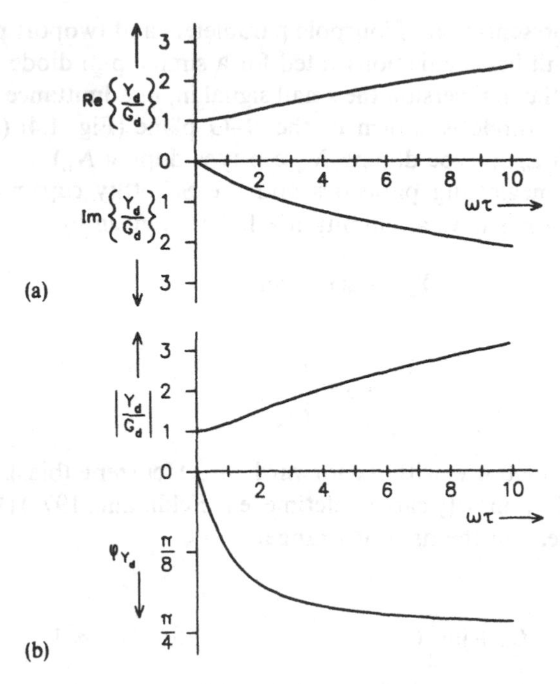

Fig. 1.5 Frequency dependence of the admittance of a p^+–n diode. (a) Real and imaginary part; (b) magnitude and phase angle.

The common complex plane B–G is mostly used for the representation of impedances at low frequencies, where the imaginary and real parts can be measured relatively simply.

A further possibility is to plot the real and imaginary part or the magnitude and phase angle separately as a function of frequency. The latter is common for stability and feedback considerations as in case of operational amplifiers, (section 4.1.1). Figure 1.5 shows these graphical representations for the same configuration as shown in Fig. 1.4.

The third version is the use of the Smith chart, where the reflection coefficient $r = |r|\,e^{j\varphi_r}$ of a device at a given input plane is plotted. This type of characterization belongs to the wave concept and is applied at higher frequencies, where voltages and currents and therefore impedances and admittances cannot be well defined or measured. The description is presented in the following paragraph.

1.3.2 Twoport parameters

At frequencies in the GHz range the complex reflection coefficients r or the S-parameters are used to characterize a device. They can be represented in a complex plane by the vector $r = |r| e^{j\varphi_r}$ or $S = |S| e^{j\varphi_S}$, which leads to the third version of the graphical impedance representation.

The parameters S_{11}, S_{22} are characteristic for the one port behaviour at input and output of the twoport, respectively, given by $S_{11} = |S_{11}| e^{j\varphi_{S_{11}}}$ and $S_{22} = |S_{22}| e^{j\varphi_{S_{22}}}$. They correspond to reflection coefficients r, (section 1.2). Therefore the Smith chart is adaptable for their representation, which consists of the complex plane $r = |r| e^{j\varphi_r}$ (Fig. 1.6).

At a given circuit plane the reflection coefficient is given by

$$r = \frac{a}{b} = \frac{(V/Z_0{}^{1/2}) - I(Z_0)^{1/2}}{(V/Z_0{}^{1/2}) + I(Z_0)^{1/2}} \tag{1.59}$$

Introducing the impedance $Z = V/I$ connected at this plane to the transmission line of the impedance Z_0 (the characteristic impedance of the measuring system)

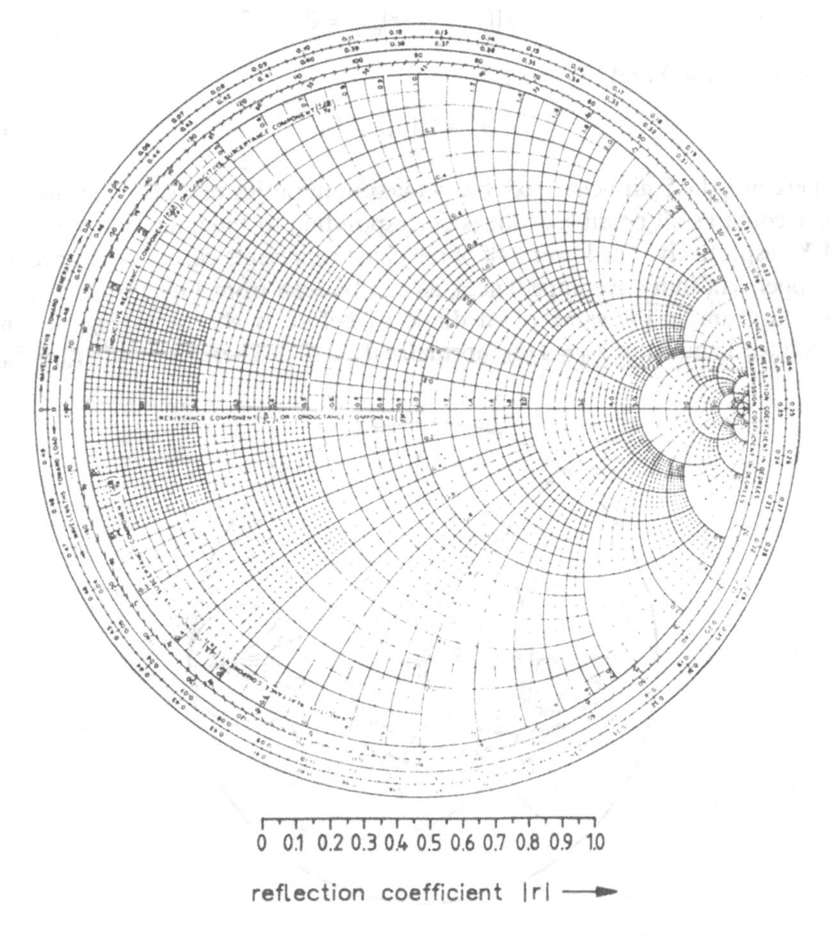

Fig. 1.6 Smith chart.

this is equivalent to

$$r = \frac{(V/I) - Z_0}{(V/I) + Z_0} = \frac{Z - Z_0}{Z + Z_0} = -\frac{Y - Y_0}{Y + Y_0} \tag{1.60}$$

$(Y = 1/Z,\ Y_0 = 1/Z_0)$.

Commonly a normalized version is used referring to the characteristic impedance Z_0 of the measuring setup, e.g. $Z_0 = 50\,\Omega$.

Taking $Z/Z_0 = z$, $Y/Y_0 = y$, it follows for the normalized version that

$$r = \frac{z-1}{z+1} = -\frac{y-1}{y+1} \tag{1.61}$$

Because for $Z = Z_0$ no reflection occurs, it is found that

$$|r||_{Z=Z_0} = |r|_{\min} = 0 \tag{1.62}$$

whereas for $Z = 0$ and $|Z| = \infty$,

$$|r| = |r|_{\max} = 1 \tag{1.63}$$

Therefore, the Smith chart exhibits a maximum radius of unity. The circle area $|r| \leqslant 1$ contains all possible r-values corresponding to the unlimited values of Z and Y, respectively. The lower half of the chart represents capacitive impedances, the upper half inductive impedances, as can be seen from Fig. 1.6.

As an example the same p^+–n diode as before is shown in Fig. 1.7 in its impedance behaviour $Z_d = 1/Y_d$ normalized to Z_0. Therefore with $z_d = Z_d/Z_0$

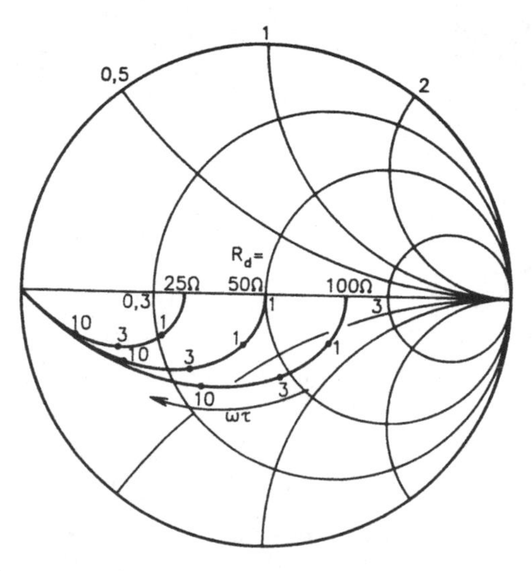

Fig. 1.7 Small signal impedance of an ideal p^+–n diode.

and $r_d = 1/G_d Z_0$ the quantity

$$r = \frac{z_d - 1}{z_d + 1} = \frac{r_d(1 + j\omega\tau)^{-1/2} - 1}{r_d(1 + j\omega\tau)^{-1/2} + 1} \tag{1.64}$$

is plotted. The curves shown belong to $R_d = 25\,\Omega$, $R_d = Z_0 = 50\,\Omega$ and $R_d = 100\,\Omega$, which correspond at room temperature ($V_T \approx 26\,\text{mV}$) to bias currents of about 1 mA, 0.5 mA and 0.25 mA, respectively; the carrier lifetime τ is assumed to be $\tau = 1\,\mu\text{s}$.

In the Smith chart the $(|r|, \varphi_r)$ values belonging to the z coordinates R/Z_0 and X/Z_0, corresponding to the equivalent impedance $Z = R + jX$, can easily be found (Fig. 1.6). This allows not only a simple conversion of an impedance into the equivalent complex reflection coefficient and *vice versa* but also the graphical transformation (inversion) of $Z \to 1/Z = Y = G + jB$ and $Y \to Z = 1/Y = R + jX$. Taking $-r$ and interpreting the coordinates R/Z_0 as G/Y_0 and X/Z_0 as B/Z_0 the inverted quantities are found. Thus the Smith chart allows us easily to achieve graphically the complex transformations $Z \Leftrightarrow Y$ as well as

$$r \to z = \frac{1 + r}{1 - r}, \; z \to r = \frac{z - 1}{z + 1}$$

and

$$r \to y = \frac{r + 1}{r - 1}, \; y \to r = \frac{1 - y}{1 + y} \tag{1.65}$$

As shown before, the scattering parameters S_{11} and S_{22} are the reflection coefficients at the input and output, if no reflection occurs on the other port,

$$\left.\begin{aligned} S_{11} &= r_1|_{r_L = 0} \\ S_{22} &= r_2|_{r_G = 0} \end{aligned}\right\} \tag{1.66}$$

As a result the representation of S_{11} and S_{22} can simply be performed in the Smith chart, equivalent to any oneport device.

If $r_L \neq 0$, $r_G \neq 0$, then

$$r_1 = S_{11} + \frac{S_{12}S_{21}r_L}{1 - S_{22}r_L}, \quad r_L = \frac{Z_L - Z_2}{Z_L + Z_2} \tag{1.67}$$

and

$$r_2 = S_{22} + \frac{S_{12}S_{21}r_G}{1 - S_{11}r_G}, \quad r_G = \frac{Z_G - Z_1}{Z_G + Z_1} \tag{1.68}$$

Again measured values r_1,r_2 can be shown in the Smith chart and directly be transformed in the corresponding impedances Z_1, Z_2, respectively.

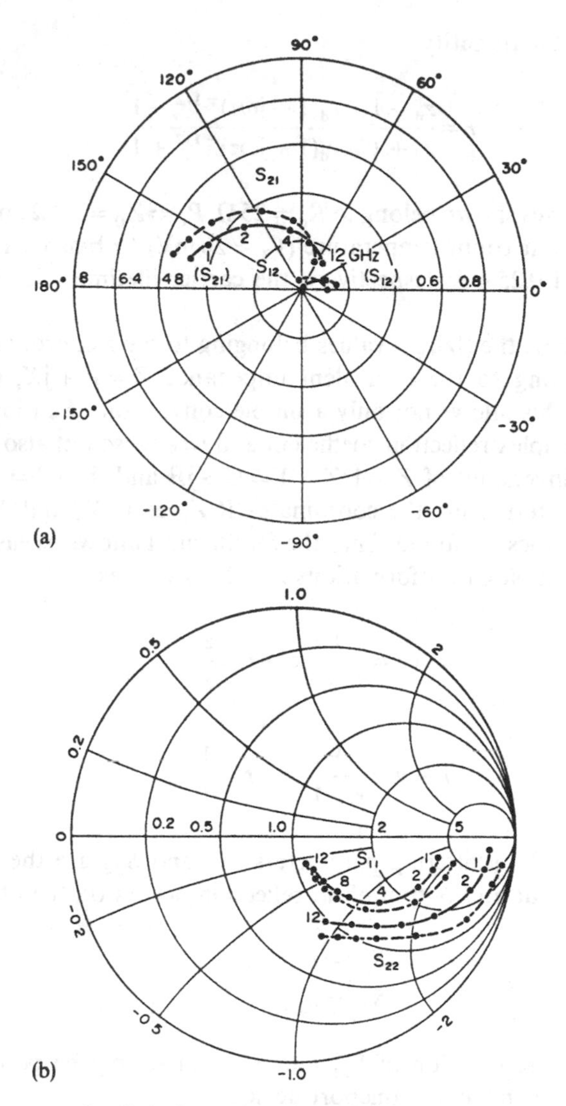

Fig. 1.8 *S*-parameters of a GaAs heterojunction bipolar transistor. (——) measured; (---) computed (SPICE-F). (After Madihan *et al.*, 1987.)

The parameters S_{12}, S_{21} are transfer parameters and therefore not related to one single port. As mentioned at the beginning of this section they can be shown in a complex plane as well. Again a Smith chart-like plane can be used, however, its amplitude scaling has to be interpreted in another way; the magnitudes $|S_{21}|$, $|S_{12}|$ of the two quantities $S_{12} = |S_{12}|e^{j\varphi S_{12}}$, $S_{21} = |S_{21}|^{j\varphi S_{21}}$ might not fit its maximum radius of $|r| = 1$ very well. Therefore the amplitude scale can be chosen corresponding to the magnitude of these quantities; generally $|S_{12}| \ll 1$ and $|S_{21}| > 1$ for common active devices.

As example, Fig. 1.8 shows *S*-parameters of a bipolar transistor, Fig. 1.9 those of a field-effect transistor and Fig. 1.10 gives *S*-parameters of a permeable base transistor amplifier.

In the last case $|S_{12}|$, $|S_{21}|$ are given in decibels, a convention often found in the literature but incorrect. From section 1.2 it follows that the squares $|S_{kl}|^2$ are power ratios, leading correctly to

$$\frac{|S_{kl}|^2}{\text{dB}} = 10 \log \{|S_{kl}|^2\} \tag{1.69}$$

However, $|S_{kl}|$ itself is not a power ratio and should not be given in decibels; '$|S_{12}| = -25$ dB' means $|S_{12}|^2 = -25$ dB and therefore $|S_{12}| = 0.056$, and '$|S_{21}| =$ 5dB' means $|S_{21}|^2 = 5$ dB and therefore $|S_{21}| = 1.78$, corresponding to the (false) interpretation

$$\frac{`|S_{kl}|\text{'}}{\text{dB}} = 20 \log \{|S_{kl}|\} \tag{1.70}$$

Only

$$\frac{|S_{kl}|^2}{\text{dB}} = 20 \log \{|S_{kl}|\} \tag{1.71}$$

can be written correctly.

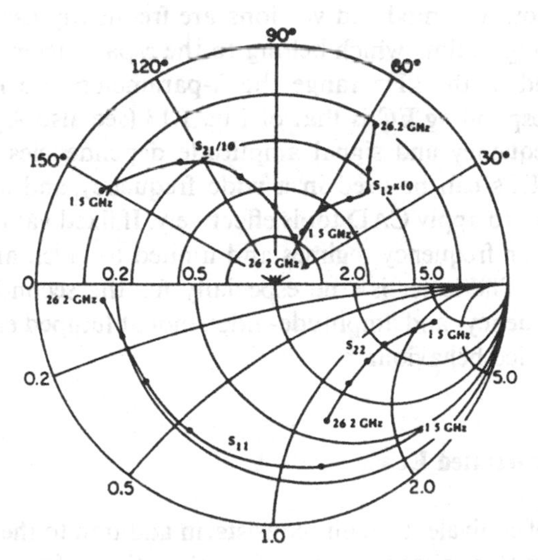

Fig. 1.9 *S*-parameters of an AlInAs/GaInAs MODFET as a function of frequency on the Smith chart. Note that in order to keep S_{21} within the circle of radius 1, it is divided by 10. Likewise S_{12} is multiplied by 10 so that its value can be accurately shown on the diagram. (After Peng *et al.*, 1987.)

1.4 EQUIVALENT CIRCUITS

Two groups of equivalent circuits (ECs) have to be distinguished; these will be considered next.

1.4.1 Parameter-related ECs

The first group consists of a visualization of the fourpole (or twoport) parameters, as shown in Appendix A. These more formal ECs exhibit components which correspond directly to the given fourpole parameter. Transfer parameters (indices 12, 21) have to be transformed into current sources and voltages, controlled by input and output signals, respectively. Four possibilities are indicated in Fig. 1.11. The current source (Fig. 1.11(a), (b)) is assumed to have zero conductance (open circuit), whereas the voltage source (Fig. 1.11(c), (d)) has zero resistance (short circuit). Either a voltage control or a current control can be used corresponding to the control function indicated in the fourpole equation (Appendix A). The arrows indicate the positive direction of current flow and voltage signal, respectively. Their directions have to be defined but they can be chosen arbitrarily.

As a result, the current–voltage relationships of these ECs represent the fourpole equations directly. Different forms are possible for the same fourpole equations, which is demonstrated in Fig. 1.12 for the *y*-parameter. Figure 1.13 shows the complete *h*-parameter equivalent circuit.

The Π-configuration after Fig. 1.12(c) is the commonly used EC in the MHz and GHz range. However, modified versions are frequently used, such as those based on the T-configuration, which belong to the *z*-parameters (section 5.3). At low frequencies and in the kHz range the *h*-parameters are mostly adopted. Therefore the corresponding EC is that of Fig. 1.13 (see also Appendix A).

If the proper frequency and signal amplitude dependencies are chosen, the parameter-related ECs can be used in a wide frequency and signal amplitude range. This allows us to apply CAD tools effectively. If fixed values are used, they are restricted to given frequency regimes and limited to given amplitudes of the applied RF signals. The latter is true especially for the second group of ECs, where constant frequency- and amplitude-independent lumped elements are used to represent the device behaviour.

1.4.2 Device-physics-related ECs

The second group of equivalent circuits consists, in addition to the implementation of mutual current and/or voltage sources, of combinations of passive components: resistors, inductors and capacitors. These ECs are restricted to given frequency regimes and limited to given amplitudes of the applied RF signals, as mentioned above. They are derived from the physical model of the device, including parasitic

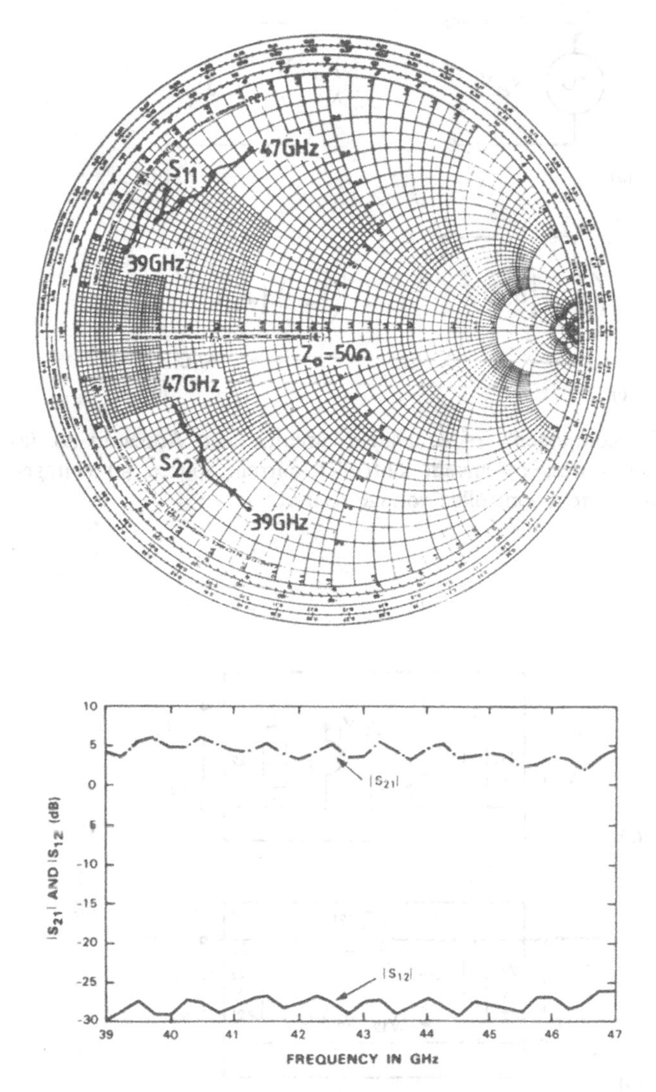

Fig. 1.10 *S*-parameters of an EHF amplifier with a GaAs permeable base transistor. (After Actis *et al.*, 1987.)

elements, and can be developed step by step. In this case a combination of different lumped elements represents the structure electrically, not directly corresponding to one parameter of the general equivalent circuit. Examples are given in Chapter 5.

Here the behaviour of a p–n diode will be treated as previously (section 1.3.1). There the small signal admittance of an ideal p^+–n diode was discussed, leading

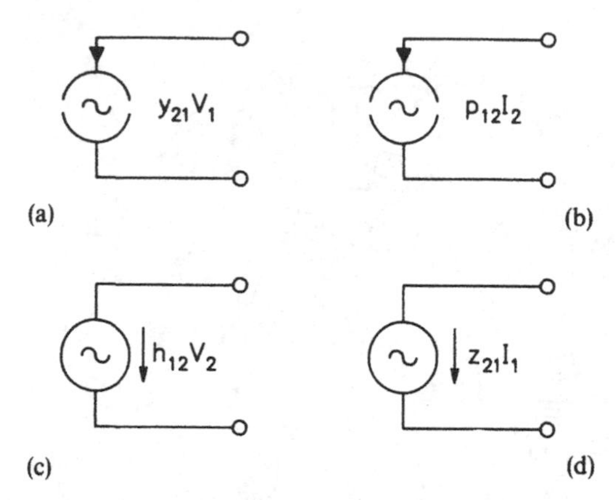

Fig. 1.11 Transfer components of equivalent circuits (also Appendix A). (a) Voltage-controlled current source; (b) current-controlled current source; (c) voltage-controlled voltage source; (d) current-controlled voltage source.

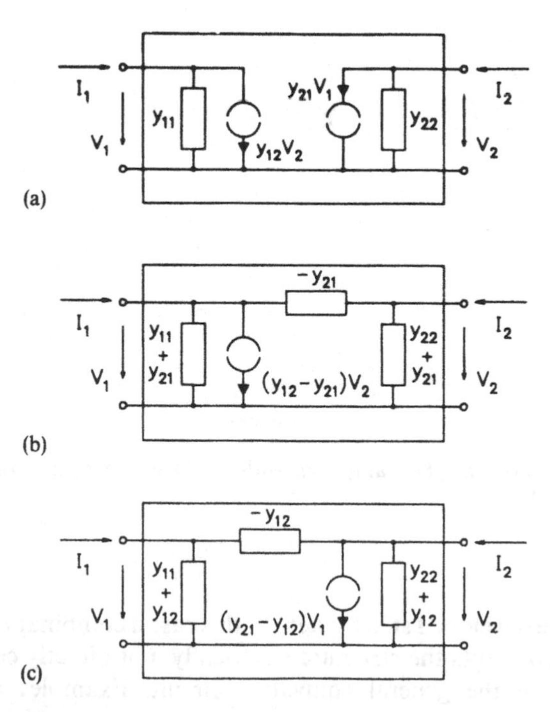

Fig. 1.12 Equivalent circuits representing the admittance matrix. (a) Separated ECs for input and output; (b) combined EC, mutual current source at the input; (c) combined EC, mutual current source at the output.

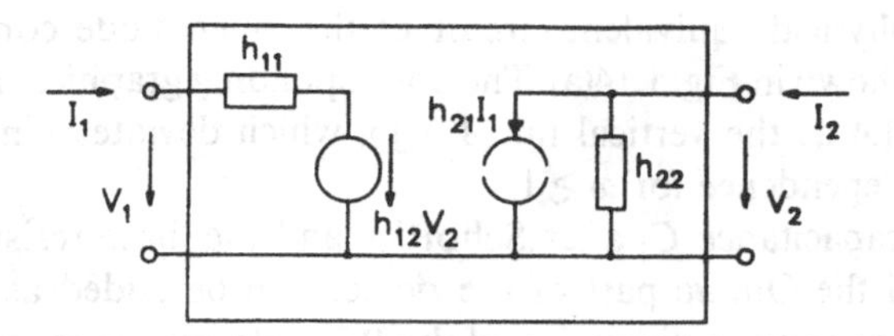

Fig. 1.13 Equivalent circuit (*h*-parameters).

in the forward direction ($I_P > 0$) to an admittance

$$Y_d = G_d(1 + j\omega\tau)^{1/2} \tag{1.72}$$

Only at low frequencies, $\omega \ll 1/\tau$, can frequency-independent components be extracted, leading to

$$Y_d = G_d + j\omega\frac{G_d\tau}{2} \tag{1.73}$$

The term

$$\frac{G_d\tau}{2} = C_d \tag{1.74}$$

represents the diffusion capacitance; G_d is the diffusion conductance of the diode.

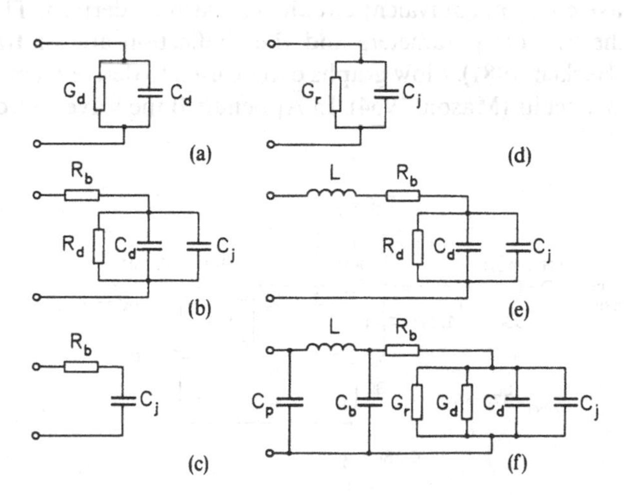

Fig. 1.14 Diode equivalent circuits. (a) Inner diode, $I_P > 0$ (forward biased). (b) diode with base resistance and junction capacitance; (c) reverse-biased diode ($I_P < 0$), at high frequencies; (d) reverse-biased diode, at low frequencies; (e) diode with bonding wire; (f) Complete EC with mounting parasitics.

Therefore, a physical equivalent circuit of the inner diode consists of G_d in parallel to C_d as shown in Fig. 1.14(a). The corresponding graphical representation is given in Fig. 1.4 as the vertical line $1 + jb$, which deviates remarkably from the true curved dependence for $\omega \gtrsim 1/\tau$.

The junction capacitance C_j after Schottky and the 'base resistance' R_b as a representation of the Ohmic part of the device can be added as the next step (Fig. 1.14(b)). This permits a discussion of the RF performance in reverse direction ($I_P < 0$), as G_d and C_d become zero. The remaining part consists of R_b and C_j (Fig. 1.14(c)), which combination is of importance for varactor diodes. In principle a dynamic reverse conductance G_r has to be implemented (Fig. 1.14(d)). However, at sufficiently high frequencies ($\omega \gg G_r/C_j$) it can be neglected. By adding an inductance L, representing e.g. a bonding wire (Fig. 1.14(e)), a more complicated EC is derived. Including at last the parasitic components of the mounting gives the final EC (Fig. 1.14(f)). Therewith most of the problems related to the RF behaviour of the device under discussion can be considered.

In conclusion, device physics-related ECs can implement the different physical effects which govern the RF behaviour. They permit us to extract dominant effects for the development of design rules. After the adoption of a given circuit configuration as the equivalent circuit, again CAD tools can be used to minimize the difference between the real behaviour and the computed one using a network analysis program. Therewith the magnitudes of the components involved can be fixed after having started with estimated values. Figure 1.8 gives an example for the case of a GaAs heterojunction bipolar transistor (HBT), where measured and calculated S-parameters are compared, the latter belonging to the equivalent circuit given in Fig. 1.15.

Using the wave concept, equivalent circuits can also be derived. They have to be related to the twoport parameters and their reflection and/or transmission behaviour (e.g. Hecken, 1981). Flow graphs can be used to derive them, applicable also for complex circuits (Mason, 1954). In Appendix B the wave source is shown

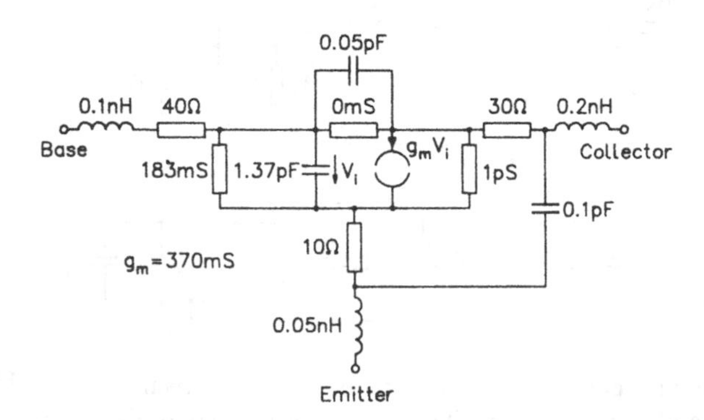

Fig. 1.15 Equivalent circuit of an HBT used for the calculation of its S-parameters shown in Fig 1.8. (3·8 μm² two-emitter HBT, $V_{CE} = 3$ V, $I_C = 14$ mA.) (After Madihan *et al.*, 1987.)

as an example which represents a reflection-free generator connected to the transmission line.

EXERCISES

1.1 Extract the small signal *h*-parameters from the DC dependencies of currents (I_1, I_2) and voltages (V_1, V_2) of a fourpole (Fig. 1.2) and determine their meaning.

1.2 Transform the hybrid parameters h_{kl} to the admittance parameters y_{kl} and *vice versa* $(k, l = 1, 2)$.

1.3 Develop the *y*-parameters y_{kl} of a sixpole configuration where the DC dependences are given with $I_1 = f_1(V_1, V_2, V_3)$, $I_2 = f_2(V_1, V_2, V_3)$ and $I_3 = f_3(V_1, V_2, V_3)$.

1.4 Express the wave parameters S_{11} and S_{21} by *z*-parameters.

1.5 Show the equivalence of the wave source and a conventional generator (Figs 1.2, 1.3, B.1).

1.6 Show the use of the *T*-parameters for sequential twoports (Fig. Ex1.1). Derive the matrix equation for the series coupling of two twoports characterized by their *T*-matrices $(\boldsymbol{T}_{\mathrm{I}}, \boldsymbol{T}_{\mathrm{II}})$.

1.7 Given a twoport connected to a load impedance exhibiting

$$r_{\mathrm{L}} = 0.6\,\mathrm{e}^{\mathrm{j}(3\pi/4)}$$

(arc $S_{11} = \pi/4$).

Determine the complex *S*-parameters S_{11}, S_{12} and S_{22} of the lossless and reciprocal twoport to achieve at its entrance port $r_1 = 0$.

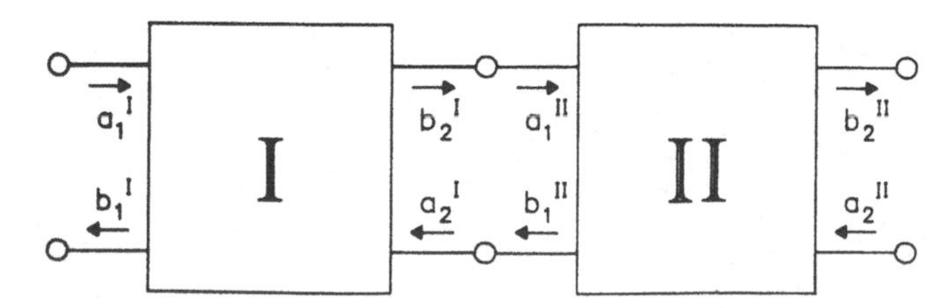

Fig. Ex1.1 Cascaded twoports, showing the relevant waves.

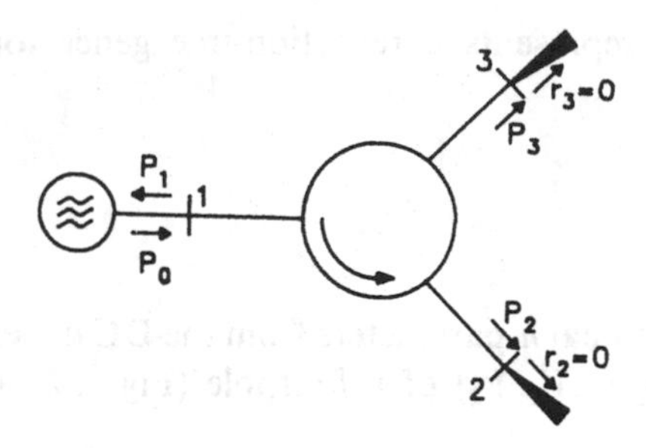

Fig. Ex1.2 Circuit with circulator including generator and matched loads.

1.8 A nonideal, symmetrical circulator exhibits a scattering matrix

$$S_c = \begin{bmatrix} \rho & \sigma & \tau \\ \tau & \rho & \sigma \\ \sigma & \tau & \rho \end{bmatrix}.$$

The circuit of Fig. Ex1.2 is used to determine $|\rho|$, $|\sigma|$, $|\tau|$. We have

$$\frac{P_1}{P_0} \mathrel{\hat{=}} -26\,\mathrm{dB}$$

(P_0 is the power of the generator),

$$\frac{P_2}{P_0} \mathrel{\hat{=}} -3\,\mathrm{dB}$$

$$\frac{P_3}{P_0} \mathrel{\hat{=}} -20\,\mathrm{dB}$$

How large are the parameters $|\rho|$, $|\sigma|$, $|\tau|$?

2

Analog small signal amplification

The small signal behaviour of semiconductor devices can be described by linearized functions (section 1.1), by applying the fourpole and twoport formulations derived in Chapter 1. If nonlinear distortions have to be considered a more detailed description is needed including higher order dependencies.

In practice the term 'small signal' is stressed somewhat and applied to voltage signals with amplitudes up to a few volts. From a theoretical point of view, however, amplitudes are small only if higher order terms of the Taylor series can be neglected (section 1.1).

If junctions and minority carrier injections are considered, applied voltage amplitudes have to be small in comparison to the thermal voltage V_T ($\approx 26\,\mathrm{mV}$ at room temperature), if the common linearized equivalent circuits are adaptable. This follows from the exponential voltage dependence of the minority carrier density at a junction, e.g.

$$p = p_n \mathrm{e}^{V/V_T} \tag{2.1}$$

where V consists of a bias value V_P and an applied sinusoidal signal $v(t)$.

This leads to

$$p = p_n \exp\left(\frac{V_P + v(t)}{V_T}\right) = p_n \mathrm{e}^{V_P/V_T} \sum_{m=0}^{\infty} \frac{1}{m!}\left(\frac{v(t)}{V_T}\right)^m \tag{2.2}$$

This expression can be linearized only for $|v(t)| \ll V_T$ leading to

$$p = p_n \mathrm{e}^{V_P/V_T}\left\{1 + \frac{v(t)}{V_T}\right\} = \bar{p} + \tilde{p} \tag{2.3}$$

The two components are the bias-dependent term

$$\bar{p} = p_n \mathrm{e}^{V_P/V_T} \tag{2.4}$$

Table 2.1 Power gain definitions

Effective gain G_{eff}	$\frac{P_L}{P_1} = \frac{\lvert V_2\rvert^2 G_L}{\lvert V_1\rvert^2 G_1}$
Insertion gain G_{IN}	$\frac{P_L}{P_L\rvert_{\text{without amplifier}}} = \frac{\lvert V_2\rvert^2 G_L}{\lvert I_G\rvert^2 G_L/(G_L+G_G)^2}$
Transducer gain G_T	$\frac{P_L}{P_{AVG}} = \frac{\lvert V_2\rvert^2 G_L}{\lvert I_G\rvert^2/4G_G}$
Available gain G_{AV}	$\frac{P_{AVL}}{P_{AVG}} = \frac{P_L\rvert_{Y_1=[Y_2(Y_G)]^*}}{\lvert I_G\rvert^2/4G_G}$
Maximum Available gain MAG	$\left.\frac{P_{AVLopt}}{P_{AVG}}\right\rvert_{\substack{Y_L=Y_2^* \\ Y_G=Y_1^*}}$
Unilateral gain U	$\text{MAG}\rvert_{\substack{\text{lossless} \\ \text{neutralized}}}$
Maximum stable gain MSG	$G_T\rvert_{K=1 \triangleq \delta+\frac{\Delta^2}{4}=1}$

and the small signal term

$$\tilde{p} = \bar{p}\frac{v(t)}{V_T} \tag{2.5}$$

Therefore real 'small signals' have to have amplitudes of below about 5 mV, to fulfil the condition $|v| \ll V_T$ (≈ 26 mV).

2.1 POWER GAIN DEFINITIONS

From Table 2.1 it can be seen that different power gain formulations are possible based on the general amplifier configuration (Fig. 2.1). They are related to different objectives which have to be understood for correct interpretation. In Table 2.2 and Table 2.3 the relationships are expressed by the y-parameters and S-parameters, respectively.

The effective power gain or simply 'power gain' G_{eff} describes the output power of the device delivered to a load in relation to the corresponding input power.

Table 2.2 Power gain definitions, expressed in y-parameters

Effective gain G_{eff}	$\dfrac{\lvert y_{21}\rvert^2 G_L}{\lvert y_{22}+Y_L\rvert^2 \operatorname{Re}\left\{y_{11}-\dfrac{y_{12}y_{21}}{y_{22}+Y_L}\right\}}$
Insertion gain G_{IN}	$\dfrac{\lvert y_{21}\rvert^2 (G_L+G_G)^2}{\lvert (y_{11}+Y_G)(y_{22}+Y_L)-y_{12}y_{21}\rvert^2}$
Transducer gain G_T	$\dfrac{4\lvert y_{21}\rvert^2 G_L G_G}{\lvert (y_{11}+Y_G)(y_{22}+Y_L)-y_{12}y_{21}\rvert^2}$
Available gain G_{AV}	$\dfrac{\lvert y_{21}\rvert^2 G_G}{\lvert y_{11}+Y_G\rvert^2 \operatorname{Re}\left\{y_{22}-\dfrac{y_{12}y_{21}}{y_{11}+Y_G}\right\}}$
Maximum available gain MAG	$\dfrac{\lvert y_{21}\rvert^2}{2g_{11}g_{22}-\operatorname{Re}\{y_{12}y_{21}\}+[(2g_{11}g_{22}-\operatorname{Re}\{y_{12}y_{21}\})^2-\lvert y_{12}y_{21}\rvert^2]^{1/2}}$
Unilateral gain U	$\dfrac{\lvert y_{21}-y_{12}\rvert^2}{4(g_{11}g_{22}-g_{12}g_{21})}$
Maximum stable gain MSG	$\left\lvert \dfrac{y_{21}}{y_{12}} \right\rvert$

Because the input admittance can be expressed by

$$Y_1 = y_{11} - \frac{y_{12}y_{21}}{y_{22}+Y_L} \tag{2.6}$$

this relationship is dependent on the load admittance.

The **insertion gain** G_{IN} is the comparison of the power delivered into a given load either connected directly to a given generator or via the amplifier, inserted between this generator and the given load. Correspondingly an **insertion loss** L can be defined by $L_{IN} = 1/G_{IN}$ if $G_{IN} < 1$ as usual for lossy passive interconnects, transmission lines, etc.

The **transducer gain** G_T is a standard expression where the power delivered to a load is compared to the power which could be delivered directly by the generator under matched condition. This expression is dependent on the load admittance as well as on the generator admittance. Only for $G_T > 1$ does it make sense to apply the amplifier, otherwise the matched direct connection load generator would be preferable.

Table 2.3 Power gain definitions, expressed in S-parameters

Effective gain G_{eff}	$\dfrac{\lvert S_{21}\rvert^2(1-\lvert\Gamma_L\rvert^2)}{(1-\lvert S_{11}\rvert^2)+\lvert\Gamma_L\rvert^2(\lvert S_{22}\rvert^2-\lvert\Delta S\rvert^2)-2\,\mathrm{Re}\{\Gamma_L N\}}$
Insertion gain G_{IN}	$\lvert S_{21}\rvert^2$
Transducer gain G_T	$\dfrac{\lvert S_{21}\rvert^2(1-\lvert\Gamma_S\rvert^2)(1-\lvert\Gamma_L\rvert^2)}{(1-S_{11}\Gamma_S)(1-S_{22}\Gamma_L)-S_{12}S_{21}\Gamma_L\Gamma_S}$
Available gain G_{AV}	$\dfrac{\lvert S_{21}\rvert^2(1-\lvert\Gamma_S\rvert^2)}{(1-\lvert S_{22}\rvert^2)+\lvert\Gamma_S\rvert^2(\lvert S_{11}\rvert^2-\lvert\Delta S\rvert^2)-2\,\mathrm{Re}\{\Gamma_S M\}}$
Maximum available gain MAG	$\left\lvert\dfrac{S_{21}}{S_{12}}\right\rvert(k^2-1)^{1/2}$
Unilateral gain U	$\dfrac{1}{2}\dfrac{\left\lvert\dfrac{S_{21}}{S_{12}}-1\right\rvert^2}{k\left\lvert\dfrac{S_{21}}{S_{12}}\right\rvert-\mathrm{Re}\left\{\dfrac{S_{21}}{S_{12}}\right\}}$, $\quad k=\dfrac{1+\lvert\Delta S\rvert^2-\lvert S_{11}\rvert^2-\lvert S_{22}\rvert^2}{2\lvert S_{12}S_{21}\rvert}$
Maximum stable gain MSG	$\left\lvert\dfrac{S_{21}}{S_{12}}\right\rvert$

$\Delta S=S_{11}S_{22}-S_{12}S_{21}$, $M=S_{11}-\Delta S S_{22}^*$, $N=S_{22}-\Delta S S_{11}^*$, $\Gamma_L=\dfrac{Y_0-Y_L}{Y_0+Y_L}$, $\Gamma_S=\dfrac{Y_0-Y_G}{Y_0+Y_G}$, $Y_0=\dfrac{1}{Z_0}$ (Z_0: characteristic impedance)

The **available gain** G_{AV} is the transducer gain in the case where the output of the amplifier is matched, $Y_L=Y_2^*$ (A^* is the conjugated quantity of A). Because its output admittance

$$Y_2=y_{22}-\frac{y_{12}y_{21}}{y_{11}+Y_G} \tag{2.7}$$

is dependent on the generator admittance Y_G, this expression is dependent on Y_G, and different values of G_{AV} are achieved for different generator admittances.

The **maximum available gain** MAG is the available gain in the case of matched input as well as matched output and is therefore the maximum gain achievable for a device with given twoport parameters. Because then the values of Y_G, Y_L are expressable by the fourpole parameters, this formula contains only the fourpole parameters or the S-parameters, respectively.

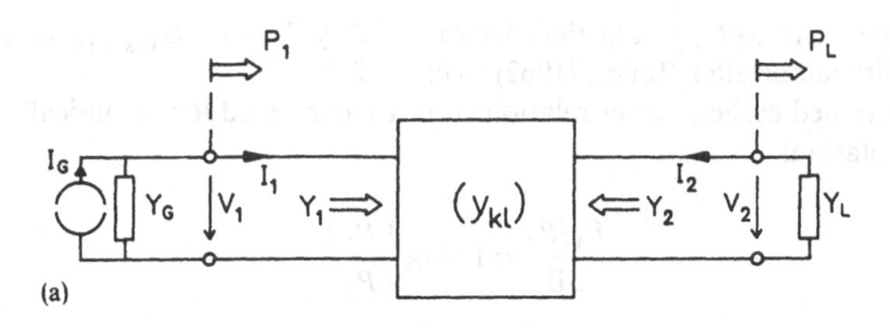

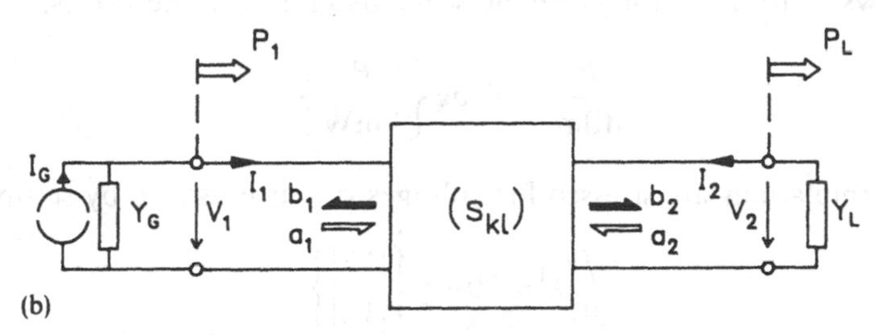

Fig. 2.1 Amplifier configurations. (a) Voltage–current concept; (b) wave concept.

However, the quantity MAG is only defined or can be measured as long as the device remains electrically stable, otherwise MAG exceeds all limits (Chapter 3). If under the optimum matching condition stability exists, a plot of MAG versus frequency is a common description of an amplifier.

The **unilateral power gain** (or unilateral gain) U after Mason (1954) also allows us to define a maximum available gain for unstable fourpoles and is a quantity well suited for the characterization of active devices in the whole frequency range of interest, including instability regimes. It is defined for a neutralized fourpole where the neutralization is achieved by lossless elements. This type of neutralization is principally possible in all cases, therefore allowing the measurement of U in the unstable region (Appendix C). The value of U, however, might be smaller than MAG in the stable region because the neutralization also cancels positive feedback incorporated in the device, therefore lowering the effective gain. It is interesting to note that the magnitude of U is independent of the grounded electrode of a three-terminal device, therefore having the same value for common emitter–source, common base–gate and emitter–source follower (common collector–drain) circuitry (Appendix C).

The **maximum stable gain** MSG is a measure to calculate approximately the potential gain of an amplifier. As shown by Rollett (1962) this quantity is invariant regarding the choice of parameters:

$$\mathrm{MSG} = \left|\frac{y_{21}}{y_{12}}\right| = \left|\frac{h_{21}}{h_{12}}\right| = \left|\frac{S_{21}}{S_{12}}\right| = \cdots \tag{2.8}$$

It is equivalent to G_T just at the edge of stability, $\mathrm{MSG} = G_T|_{K=1}$, where K is the stability factor after Rollett (1962) (section 2.2).

As mentioned earlier, power relationships are expressed logarithmically in the decibel notation:

$$\frac{P_1/P_2}{\mathrm{dB}} = 10\log\left\{\frac{P_1}{P_2}\right\} \tag{2.9}$$

The absolute power $P = P_m = 1\,\mathrm{mW}$ is defined to be equivalent to $0\,\mathrm{dBm}$, which allows us to use the logarithmic scale also for absolute values,

$$\frac{P}{\mathrm{dBm}} = 10\log\left\{\frac{P}{1\,\mathrm{mW}}\right\} \tag{2.10}$$

These formulations are stressed for voltages or currents also by setting

$$\frac{|V_1/V_2|}{\mathrm{dB}} = 20\log\left\{\left|\frac{V_1}{V_2}\right|\right\} \tag{2.11}$$

$$\frac{|I_1/I_2|}{\mathrm{dB}} = 20\log\left\{\left|\frac{I_1}{I_2}\right|\right\} \tag{2.12}$$

and also

$$\frac{|V|}{\mathrm{dBV}} = 20\log\left\{\left|\frac{V}{\mathrm{Volt}}\right|\right\} \tag{2.13}$$

However, this is not recommendable because the decibel is a measure of power ratio. With

$$P_1 = \frac{|V_1|^2}{2R_1}, \quad P_2 = \frac{|V_2|^2}{2R_2} \tag{2.14}$$

R_1 and R_2 being the resistances where the voltages V_1, V_2 are applied, respectively, it follows that

$$\frac{P_1}{P_2} = \left|\frac{V_1^2}{V_2^2}\right|\frac{R_2}{R_1} \tag{2.15}$$

Therefore the correct expression is

$$\frac{P_1/P_2}{\mathrm{dB}} = 20\log\left\{\left|\frac{V_1}{V_2}\right|\right\} + 10\log\left\{\frac{R_2}{R_1}\right\} \tag{2.16}$$

The second term is only zero if the corresponding resistances R_1, R_2 are equal.

2.2 MATCHING AND STABILITY

To introduce the influence of mismatching, matching and the relevance of electric stability or potential instability, the transducer gain G_T of an active device shall be discussed in admittance parameter notation (Beneking, 1966).

With

$$G_T = \frac{P_L}{P_{AVG}} \tag{2.17}$$

where P_{AVG} is the available power of the generator and P_L the power delivered to the load,

$$P_{AVG} = \left|\frac{I_G}{2^{1/2}}\right|^2 \frac{1}{4G_G} \quad \text{and} \quad P_L = \left|\frac{V_2}{2^{1/2}}\right|^2 G_L \tag{2.18}$$

it follows that

$$\begin{aligned} G_T &= 4G_G G_L \left|\frac{V_2}{I_G}\right|^2 \\ &= 4G_G G_L \frac{|y_{21}|^2}{|y_{12}y_{21} - (y_{11} + Y_G)(y_{22} + Y_L)|^2} \end{aligned} \tag{2.19}$$

If no internal feedback exists, $y_{12} = 0$, for matched input $Y_G = y_{11}^* = g_{11} - jb_{11}$ and matched output $Y_1 = y_{22}^* = g_{22} - jb_{22}$ it follows that

$$G_T|_{y_{12}=0,\text{matched}} = G_{TO} = \frac{|y_{21}|^2}{4g_{11}g_{22}} \tag{2.20}$$

which is equal to the maximum available gain in the case where $y_{12} = 0$. This expression allows a rough estimate of the available power gain of the device under consideration, assuming no inner feedback ($y_{12} = 0$), and consists of the product of half the short-circuit current gain $|G_{Is}|$ times half the open-circuit voltage gain $|G_{Vo}|$ of the device,

$$G_{TO} = \frac{|G_{Is}|}{2}\frac{|G_{Vo}|}{2} \tag{2.21}$$

Introducing the complex quantities

$$e = \frac{1}{g_{11}}(Y_G + jb_{11}) = e_r + je_i$$

$$\boldsymbol{a} = \frac{1}{g_{22}}(Y_L + jb_{22}) = a_r + ja_i \tag{2.22}$$

$$\boldsymbol{d} = \frac{y_{12}y_{21}}{g_{11}g_{22}} = \delta + j\Delta = de^{j\varphi_d}$$

G_T can be expressed in a normalized version:

$$g_T = \frac{G_T}{G_{TO}} = \frac{16e_r a_r}{|(1+\boldsymbol{e})(1+\boldsymbol{a}) - \boldsymbol{d}|^2} \tag{2.23}$$

In this equation the quantities $\boldsymbol{e}, \boldsymbol{a}$, represent the input and output, respectively, whereas $\boldsymbol{d}$ is a complex stability quantity depending on the fourpole behaviour alone. Therewith the optimum values $\boldsymbol{e}_{opt}, \boldsymbol{a}_{opt}$ and correspondingly Y_{Gopt}, Y_{Lopt} for maximum gain can be derived (Beneking, 1966).

It follows that

$$\left.\begin{aligned} e_{r,opt} &= a_{r,opt} = \left[1 - \left(\delta + \frac{\Delta^2}{4}\right)\right]^{1/2} \\ e_{i,opt} &= a_{i,opt} = \frac{\Delta}{2} \end{aligned}\right\} \tag{2.24}$$

The resulting gain becomes

$$\begin{aligned} G_{Topt} &= G_T|_{\boldsymbol{e}_{opt}, \boldsymbol{a}_{opt}} = \text{MAG} \\ &= G_{TO} \frac{4}{2\left\{1 + \left[1 - \left(\delta + \frac{\Delta^2}{4}\right)\right]^{1/2}\right\} - \delta} \end{aligned} \tag{2.25}$$

which might be larger or smaller than G_{TO} dependent on the inner feedback behaviour of the device; the smaller the denominator the higher the gain.

It can easily be seen from the normalized equation, that instability with $G_T \to \infty$ occurs for

$$(1+\boldsymbol{e})(1+\boldsymbol{a}) = \boldsymbol{d} \tag{2.26}$$

To clarify the situation the complex $\boldsymbol{d}$-plane will be considered (Fig. 2.2). Because for all positive values of e_r, a_r corresponding to passive conductances on both ports the vectors $(1+\boldsymbol{e})$ and $(1+\boldsymbol{a})$ are located in the complex plane rightwards from the straight line $(1 + j\Omega)$ with arbitrary Ω, their product $(1+\boldsymbol{e})(1+\boldsymbol{a})$ never touches area *I* in Fig. 2.2.

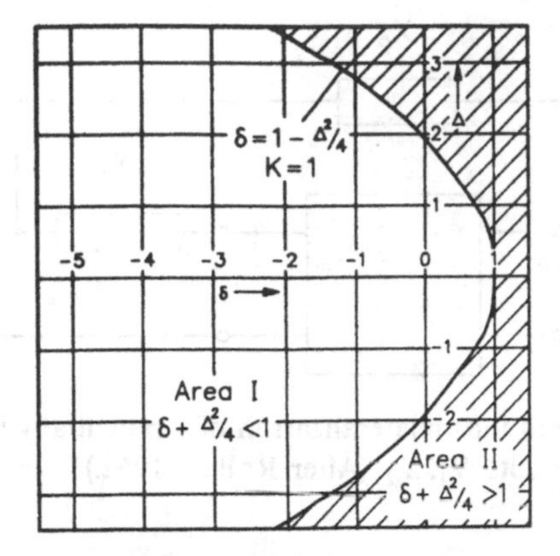

Fig. 2.2 Complex plane of the stability factor $\boldsymbol{d} = \delta + j\Delta$. For $\boldsymbol{d} \in \delta + (\Delta^2/4) < 1$ the fourpole is absolutely stable for arbitrary generator and load admittances (area I), whereas for $\boldsymbol{d} \in \delta + (\Delta^2/4) > 1$ conditional stability exists (area II). The K-factor is $K > 1$ in area I and $K < 1$ in area II. (After Beneking, 1966.)

The border line corresponds to $\delta + (\Delta^2/4) = 1$ dividing the complex plane into two parts. If the complex quantity $\boldsymbol{d}$ is located in area I, corresponding to $\delta + \Delta^2/4 < 1$ then $(1 + \boldsymbol{e})(1 + \boldsymbol{a})$ can never reach the same value whatever the outside admittances will be. Therefore a fourpole which exhibits a $\boldsymbol{d}$ value in this area I is unconditionally stable. Optimum values $\boldsymbol{e}_{\text{opt}}, \boldsymbol{a}_{\text{opt}}$ for maximum gain can be determined.

On the other hand if $\boldsymbol{d}$ is located in area II, which corresponds to $\delta + (\Delta^2/4) > 1$, at least one realizable combination $\boldsymbol{e}, \boldsymbol{a}$ exists, where $(1 + \boldsymbol{e})(1 + \boldsymbol{a})$ becomes equal to $\boldsymbol{d}$. Then instability occurs, but exclusively at this critical point. If $(1 + \boldsymbol{e})(1 + \boldsymbol{a})$ is only nearby the critical point then this circuit remains stable although the gain might be exceptionally high. Therefore the conclusion is that for $\boldsymbol{d}$ in area II the fourpole is conditionally stable, because a proper choice of $\boldsymbol{e}$ and $\boldsymbol{a}$ permits to apply this fourpole in a stable manner.

Besides the complex stability factor $\boldsymbol{d}$ other definitions are used, e.g. the Rollett factor (Rollett, 1962)

$$K = \frac{2 - \delta}{|\boldsymbol{d}|} = \frac{2g_{11}g_{22} - \operatorname{Re}\{y_{12}y_{21}\}}{|y_{12}y_{21}|} \tag{2.27}$$

If other fourpole parameters are used, the same formulation follows. As in the case of MSG this quantity is independent of the fourpole parameters used.

As long as an absolute stable configuration is considered, $K > 1$, MAG can be defined, and K as well as $\boldsymbol{d}$ can be introduced in the MAG formula. This results in

$$\text{MAG} = \text{MSG}\frac{1}{K + (K^2 - 1)^{1/2}} = \text{MSG}[K - (K^2 - 1)^{1/2}] \tag{2.28}$$

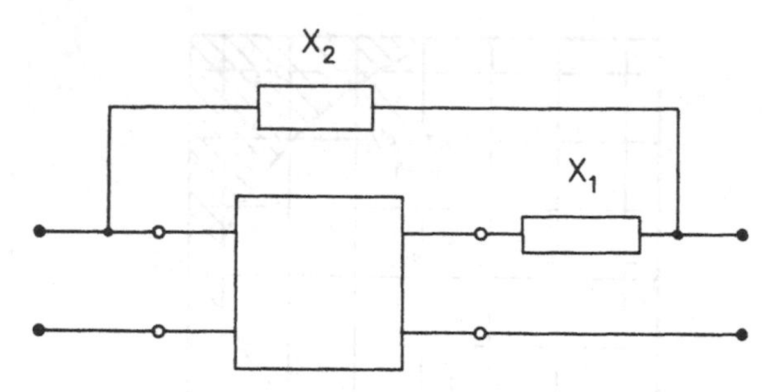

Fig. 2.3 Circuit arrangement for unilateralization of a potentially unstable fourpole with two lossless reactance elements X_1, X_2. (After Rollett, 1962.)

or

$$\mathrm{MAG} = \mathrm{MSG}\frac{|\boldsymbol{d}|}{2\left\{1+\left[1-\left(\delta+\dfrac{\Delta^2}{4}\right)\right]^{1/2}\right\}-\delta} \tag{2.29}$$

In the case of conditional stability $-1 < K < 1$, corresponding to area II in Fig. 2.2.

As can be concluded from this formula, MAG is always smaller than MSG,

$$\mathrm{MAG} \leqslant \mathrm{MSG} = \left|\frac{y_{21}}{y_{12}}\right| \tag{2.30}$$

and becomes equal to it only at the stability limit $K = 1$; this explains the notation 'maximum' stable gain MSG.

For conditionally stable fourpoles, area II of the complex $\boldsymbol{d}$-plane, with $\delta + (\Delta^2/4) > 1$, $K < 1$.

Because the distance between the complex vectors $\{(1+\boldsymbol{e})(1+\boldsymbol{a})\}$ and $\{\boldsymbol{d}\}$ in the complex $\boldsymbol{d}$-plane is responsible for the conditional stability of the given fourpole with its input and output admittances (e.g. the expression for g_T), it follows that this distance can be influenced by changing the input and output admittances, but also by changing $\boldsymbol{d}$ of the original fourpole. The latter can be achieved by partial or exact neutralization. Exact neutralization $\boldsymbol{d} = 0$ for the modified circuit, whereas partial neutralization corresponds to a sufficient change of $\boldsymbol{d}$ to avoid self-oscillation under the given conditions.

At least in principle, a complete neutralization is possible, by using only lossless elements. Figure 2.3 shows a circuit arrangement (Rollett, 1965) (also Appendix C). Therefore the unilateral gain U cannot only be computed but also measured in the instability region.

Figure 2.4 gives a comparison of U, MAG and MSG for a field-effect transistor

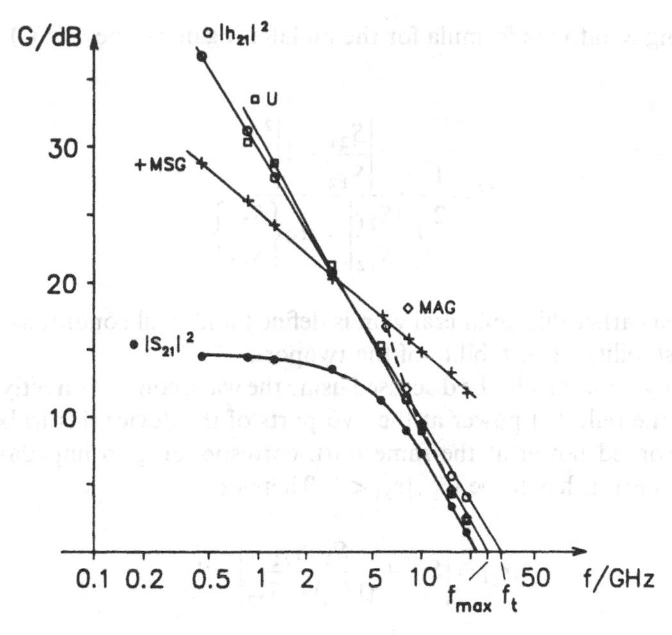

Fig. 2.4 Comparison of MAG, U, MSG and $|h_{21}|^2$ including $|S_{21}|^2$ of a field-effect transistor (NEC NE71000, one gate pad contacted).

derived from S-parameter measurements. The given device is potentially instable below $\sim 5\,\text{GHz}$, therefore MAG can be drawn only at higher frequencies.

2.3 WAVE PARAMETERS

Because

$$\text{MSG} = \left|\frac{y_{21}}{y_{12}}\right| = \left|\frac{S_{21}}{S_{12}}\right| \tag{2.31}$$

it follows that

$$\text{MAG} = \left|\frac{S_{21}}{S_{12}}\right| [K - (K^2 - 1)^{1/2}] \tag{2.32}$$

Expressed by S-parameters, K becomes

$$K = \frac{1 + |\Delta S|^2 - |S_{11}|^2 - |S_{22}|^2}{2|S_{21}S_{12}|} \tag{2.33}$$

where $\Delta S = S_{11}S_{22} - S_{12}S_{21}$.

Introducing K into the formula for the unilateral gain (Appendix C) it follows that

$$U = \frac{1}{2} \frac{\left|\dfrac{S_{21}}{S_{12}} - 1\right|^2}{K\left|\dfrac{S_{21}}{S_{12}}\right| - \mathrm{Re}\left\{\dfrac{S_{21}}{S_{12}}\right\}} \tag{2.34}$$

As mentioned earlier this unilateral gain is defined under all conditions, independent of the stability or instability of the twoport.

The stability limit can also be discussed using the wave concept directly. Because for stability the reflected power at the two ports of the device has to be smaller than the absorbed power at the same port, corresponding to impedances with positive real part, it has to be $|r_1|, |r_2| < 1$. Therefore

$$|r_1| = \left|S_{11} + \frac{S_{12}S_{21}}{(1/r_L) - S_{22}}\right| = 1 \tag{2.35}$$

and

$$|r_2| = \left|S_{22} + \frac{S_{12}S_{21}}{(1/r_G) - S_{11}}\right| = 1 \tag{2.36}$$

are the limits of stability. The expressions $r_1(r_L)$ and $r_2(r_G)$ can be considered as

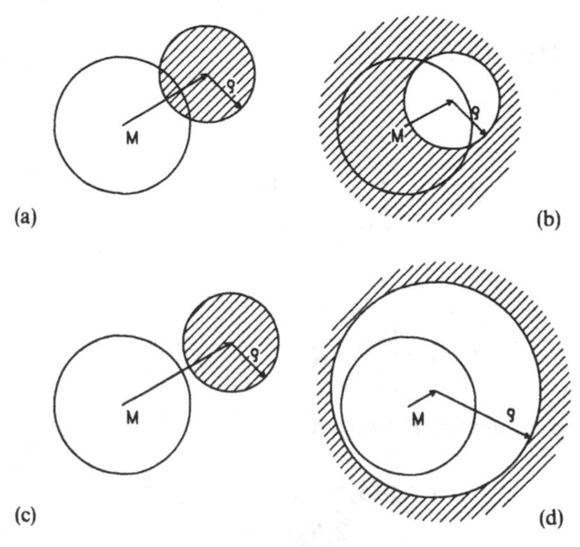

Fig. 2.5 Stability circles and Smith chart ((a), (b)) Conditional stability; ((c), (d)) unconditional stability. (After Unger and Harth, 1972.)

complex transformations leading e.g. in the r_L-plane (Smith chart) to circles corresponding to $|r_1| = 1$, the stability limit. Their centres are located at

$$M = \frac{(S_{22} - S_{11}^* \Delta S)^*}{|S_{22}|^2 - |\Delta S|^2} \tag{2.37}$$

and the radius is (Fig. 2.5)

$$\rho = \frac{|S_{12} S_{21}|}{||S_{22}|^2 - |\Delta S|^2|} \tag{2.38}$$

2.4 GRAPHICAL REPRESENTATION

The Smith chart allows not only the graphical representation of impedances but also of power relationships.

The stability situation shall be discussed first as influenced by generator and/or load impedance. Referring to the previous section four typical situations regarding the stability areas in the r_G-plane might occur, (Fig. 2.5) ($|S_{11}|, |S_{22}| < 1$). The white circle areas are the r_L-plane (Smith chart). The dashed areas correspond to $|r_1| > 1$ and characterize r_L-values where potential instability exists. Because r_L is located in the white circle area for all passive loads, configurations (c) and (d) characterize absolutely stable twoports, whereas cases (a) and (b) belong to conditionally stable twoports (Unger, 1972).

Besides this stability representation the gain of the twoport can also be shown in the Smith chart.

First the simple configuration of Fig. 2.6 (a) will be considered, where a generator with the impedance Z_G is connected to the lossless transmission line of the characteristic impedance Z_0, connected without reflection to a real load $Z_L = Z_0$. The available power of the generator is

$$P_{AV} = \frac{|V_G|^2}{2} \frac{1}{4\mathrm{Re}\{Z_G\}} \tag{2.39}$$

The power delivered to the load is

$$P_L = \frac{|V_G|^2}{2} \frac{Z_0}{|Z_0 + Z_G|^2} \tag{2.40}$$

The ratio

$$p = \frac{P_L}{P_{AV}} = \frac{4 Z_0 \,\mathrm{Re}\{Z_G\}}{|Z_0 + Z_G|^2} \tag{2.41}$$

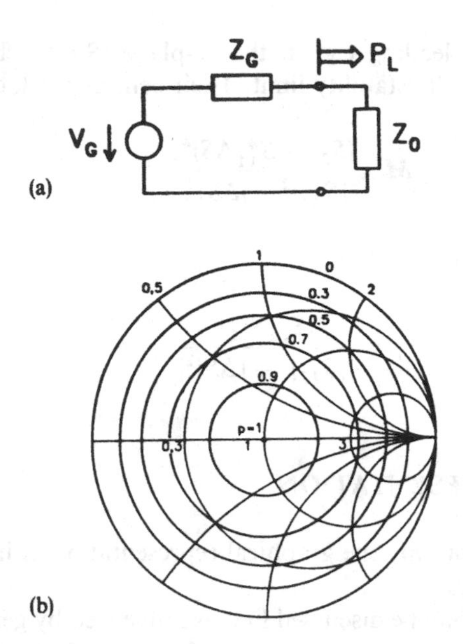

Fig. 2.6 Generator and load, (a) Equivalent circuit (Z_0 is the characteristic impedance of the transmission line with nonreflecting termination); (b) representation in the complex plane of the reflection factor r_G (parameter $p = P_L/P_{AV}$).

expressed by the reflection coefficient

$$r_G = \frac{Z_G - Z_0}{Z_G + Z_0} \tag{2.42}$$

becomes

$$p = 1 - |r_G|^2 \tag{2.43}$$

corresponding to the wave concept introduced in section 1.2. Therefore loci with $|r_G| = \text{const.}$ represent a constant power output in the Smith chart, leading to concentric circles around $r_G = 0$, with radii $|r_G| = (1 - p)^{1/2}$ for given power transfer p, (Fig. 2.6(b)).

As can be concluded from Fig. 2.6(b), impedance matching is not too critical regarding power transfer. This is also true in the case of the power gain of a fourpole in respect of the matching condition at input and output.

In the case of a twoport the loci of constant power gain are also circles in the r_G plane; however, their central points and radii are complicated functions of the fourpole/twoport parameters. As first shown by Fukui (1966), the centre points of the circles

$$r_p = |r_p| e^{-j\varphi_p} \tag{2.44}$$

are located on the same straight line. The centres are located at the points

$$r_{\mathrm{p}} = \frac{\{(1 - |y_{\mathrm{Gopt}}|^2)^2 + 4b_{\mathrm{Gopt}}\}^{1/2}}{(1 - g_{\mathrm{Gopt}})^2 + b_{\mathrm{Gopt}}^2 + 2\delta_{\mathrm{G}}} \tag{2.45}$$

where

$$y = YZ_0 = g + \mathrm{j}b \tag{2.46}$$

and

$$\delta_{\mathrm{G}} = \frac{|y_{21}|^2}{2g_{22}} Z_0 \left(\frac{1}{G_{\mathrm{AV}}} - \frac{1}{G_{\mathrm{AV\,max}}} \right) \tag{2.47}$$

The angle of the line is

$$\varphi_{\mathrm{p}} = \arctan\left(\frac{2b_{\mathrm{Gopt}}}{1 - |y_{\mathrm{Gopt}}|^2} \right) \tag{2.48}$$

Figure 2.7 gives an example of these Apollonios circles. Correspondingly the radii

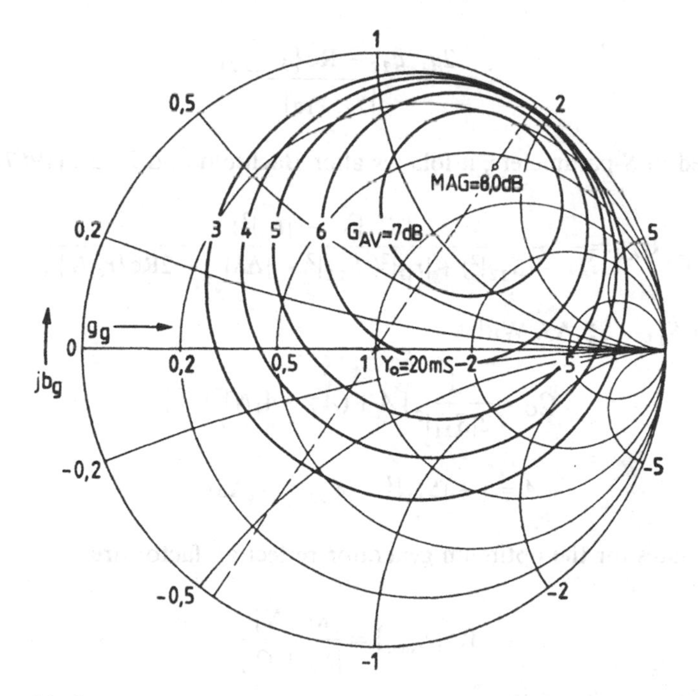

Fig. 2.7 Graphical representation of the available power gain G_{AV} in the complex r_{G} plane (Smith chart). Bipolar transistor GM 1233 (Texas Instruments; $f = 2.4\,\mathrm{GHz}$, $V_{\mathrm{CE}} = -8\,\mathrm{V}$, $I_{\mathrm{C}} = -1.5\,\mathrm{mA}$). (After Bächfold and Strutt, 1967.)

are

$$\rho_{\mathrm{p}} = \frac{2g_{\mathrm{RG}}}{(1 - g_{\mathrm{Gopt}})^2 + b_{\mathrm{Gopt}}{}^2 + 2\delta_{\mathrm{G}}} \tag{2.49}$$

where

$$g_{\mathrm{RG}} = Z_0 \left\{ \frac{G_{\mathrm{Gopt}} |y_{21}|^2}{g_{22}} \left(\frac{1}{G_{\mathrm{AV}}} - \frac{1}{G_{\mathrm{AVmax}}} \right) + \frac{|y_{21}|^4}{4g_{22}{}^2} \left(\frac{1}{G_{\mathrm{AV}}} - \frac{1}{G_{\mathrm{AVmax}}} \right)^2 \right\}^{1/2} \tag{2.50}$$

The quantities g_{Gopt} and b_{Gopt} belong to $G_{\mathrm{AVmax}} = \mathrm{MAG}$ and were defined earlier:

$$g_{\mathrm{Gopt}} = Z_0 \frac{|y_{12} y_{21}|}{2g_{22}} (K^2 - 1)^{1/2} \tag{2.51}$$

$$b_{\mathrm{Gopt}} = Z_0 \left[\frac{\mathrm{Im}\{y_{12} y_{21}\}}{2g_{22}} - b_{11} \right] \tag{2.52}$$

and

$$K = \frac{2g_{11} g_{22} - \mathrm{Re}\{y_{12} y_{21}\}}{|y_{21} y_{12}|} \tag{2.53}$$

If expressed in S-parameters, it follows after Bächtold and Strutt (1967) that

$$G_{\mathrm{AV}} = \frac{|S_{21}|^2 (1 - |r_{\mathrm{G}}|^2)}{(1 - |S_{22}|^2) + |r_{\mathrm{G}}|^2 (|S_{11}|^2 - |\Delta S|^2) - 2\mathrm{Re}\{r_{\mathrm{G}} N\}} \tag{2.54}$$

where $N = S_{11} - S_{22}^* \Delta S$. With

$$Q_{\mathrm{G}} = \frac{1}{2|S_{21}|^2} [A + (A^2 - 4|N|^2)^{1/2}] \tag{2.55}$$

$$A = 1 + |S_{11}|^2 - |S_{22}|^2 - |\Delta S|^2 \tag{2.56}$$

the coordinates for the optimum generator reflection factor are

$$\mathrm{Re}\{r_{\mathrm{Gopt}}\} = \frac{\mathrm{Re}\{N\}}{|S_{21}|^2 Q_{\mathrm{G}}} \tag{257}$$

$$\mathrm{Im}\{r_{\mathrm{Gopt}}\} = \frac{\mathrm{Im}\{N\}}{|S_{21}|^2 Q_{\mathrm{G}}} \tag{2.58}$$

therefore

$$r_{\mathrm{Gopt}} = \frac{N}{|S_{21}|^2 Q_{\mathrm{G}}} \tag{2.59}$$

EXERCISES

2.1 Develop the formula for the power gain $G_{\mathrm{eff}}(a_2)$ of a fourpole with real parameters depending on the feedback parameter f. Use the following notations:

- matching coefficient of the load resistance R_{L} at the output

$$a_2 = \frac{R_{\mathrm{L}}}{Z_{\mathrm{o2}}}$$

where Z_{o2} is the characteristic impedance at the output,

$$Z_{\mathrm{o2}} = \sqrt{R_{2\mathrm{s}} R_{2\mathrm{o}}}$$

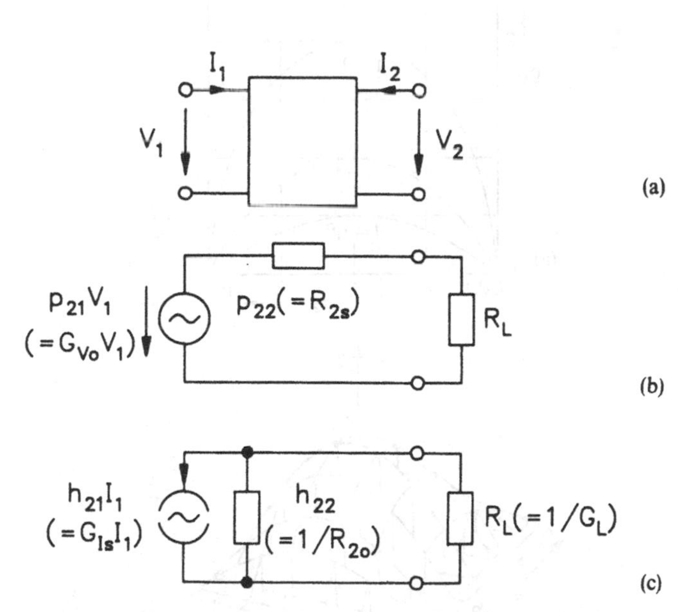

Fig. Ex2.1 Output equivalent circuits of a fourpole for the evaluation of its voltage gain G_{V} and current gain G_{I} ($G_{V_{\mathrm{o}}}$ open circuit voltage gain, $R_{\mathrm{L}} \to \infty$; G_{Is} short-circuit current gain, $R_{\mathrm{L}} \to 0$). (a) Fourpole; (b) equivalent circuit with p-parameters; (c) equivalent circuit with h-parameters.

- output resistance for short-circuit input

$$R_{2s} = \frac{1}{y_{11}}$$

- output resistance for open input

$$R_{2o} = \frac{1}{h_{22}}$$

- feedback parameter

$$f = \left(\frac{R_{2s}}{R_{2o}}\right)^{1/2}$$

First draw the equivalent output circuit for the voltage gain G_V and the current gain G_I of the fourpole (use the p-parameters and h-parameters, respectively, and combine them to achieve the power gain; the equivalent circuits can be found in Fig. Ex2.1).

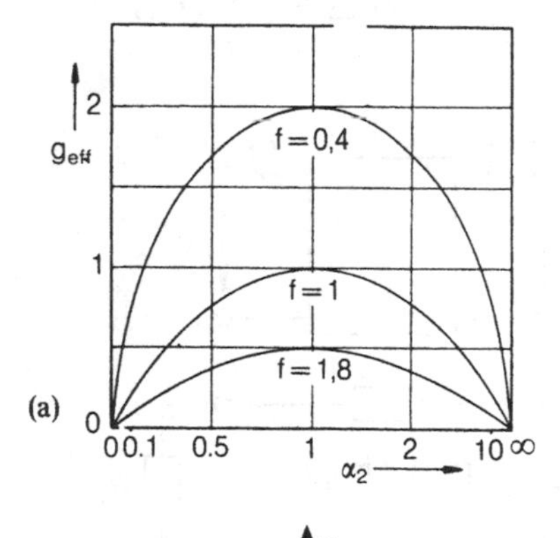

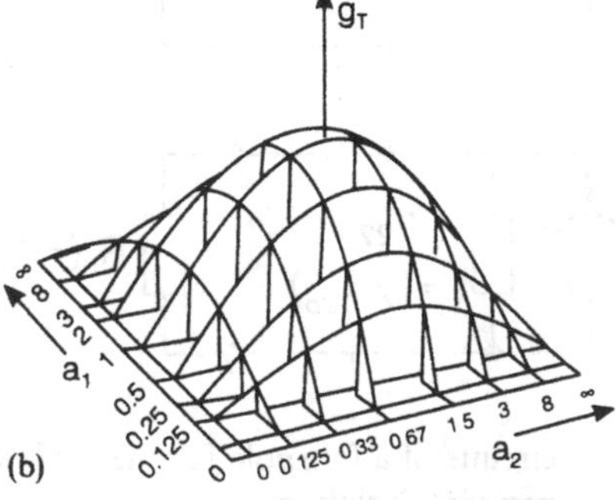

Fig. Ex2.2 Graphs of the normalized power gains. (a) $g_{eff}(a_2)$ for different values of f; (b) $g_T(a_1, a_2)$ for $f = 1$.

Show first that the voltage gain G_V and the current gain G_I are expressed by

$$G_V = \frac{G_{Vo}}{1 + (R_{2s}/R_L)}, \quad G_I = \frac{G_{Is}}{1 + (R_L/R_{2o})}$$

Combine the two expressions to achieve the power gain $G_{eff}(a_2, f)$.

Show that without feedback ($f = 1$) for $a_2 = 1$ the optimum gain is

$$G_o = \frac{G_{Vo} G_{Is}}{4}$$

Normalize G_{eff} to G_o and draw the dependence of

$$g_{eff} = \frac{G_{eff}}{G_o} = \frac{4}{1 + f[a_2 + (1/a_2)] + f^2}$$

for $f = 0.4$, $f = 1.0$ and $f = 1.8$ (Fig. Ex2.2(a)). To get symmetrical curves, for $0 \leqslant a_2 < 1$ use a linear scale, for $1 \leqslant a_2 < \infty$ use $1/a_2$.

Introduce the input matching factor

$$a_1 = \frac{R_G}{Z_1(R_L)}$$

in the formula for the normalized effective gain to achieve the normalized transducer gain

$$g_T = \frac{4}{1 + (a_1 + (1/a_1)) + 1} g_{eff}$$

where R_G is the generator resistance, connected to the input with its input impedance

$$Z_1(R_L) = z_{11} - \frac{z_{12} z_{21}}{z_{22} + R_L}$$

Draw the three-dimensional dependence $g_T(a_1, a_2)$ for $f = 1$ (Fig. Ex2.2(b)) and discuss the influence of mismatch at input and output.

Show that for mismatch in the limits of

$$0.5 \lesssim a_1, a_2 \lesssim 2$$

the power gain is not reduced too much: at the indicated limits (worst case)

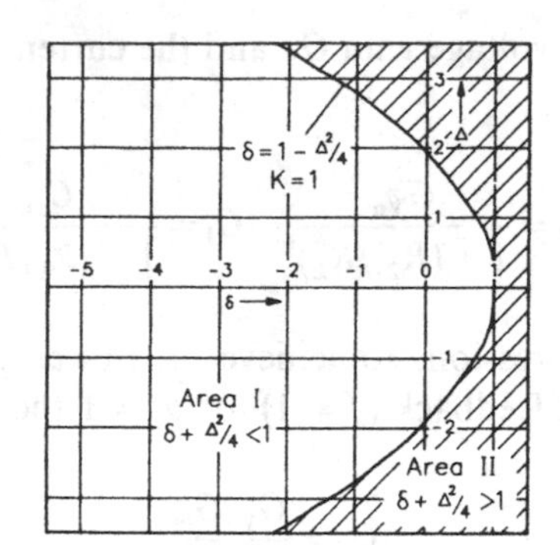

Fig. Ex2.3 Complex plane of the stability factor $d = \delta + j\Delta$ (section 2.2).

the reduction is about 20%. The conclusion is that impedance matching is not the most important issue (Fig. Ex2.2).

2.2 Show that the difference of h_{22} and y_{22} or their ratio, respectively, indicates the inner feedback behaviour of a fourpole and its stability (quasistatic approach (also exercise 2.1)).

2.3 Show in the complex $\boldsymbol{d}$ plane (Fig. Ex2.3, after Fig. 2.2) the slope of $\boldsymbol{d}$ of an

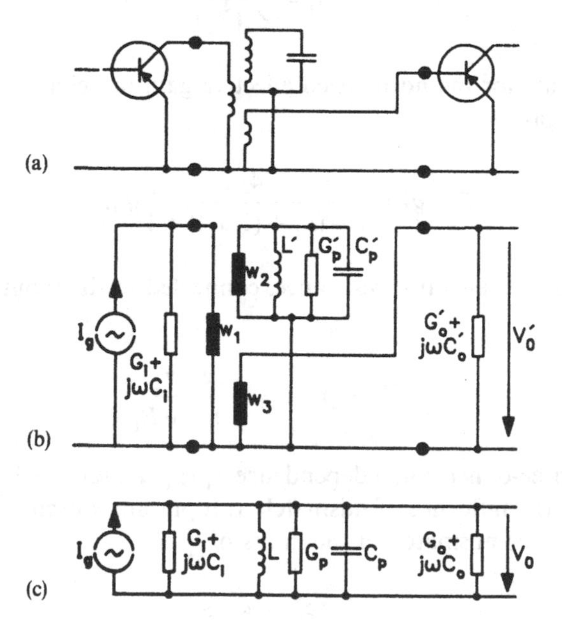

Fig. Ex2.4 Narrowband amplifier stage. (a) Circuit with matching transformer and resonant circuit; (b) equivalent circuit; (c) output related equivalent circuit. (After Beneking, 1959.)

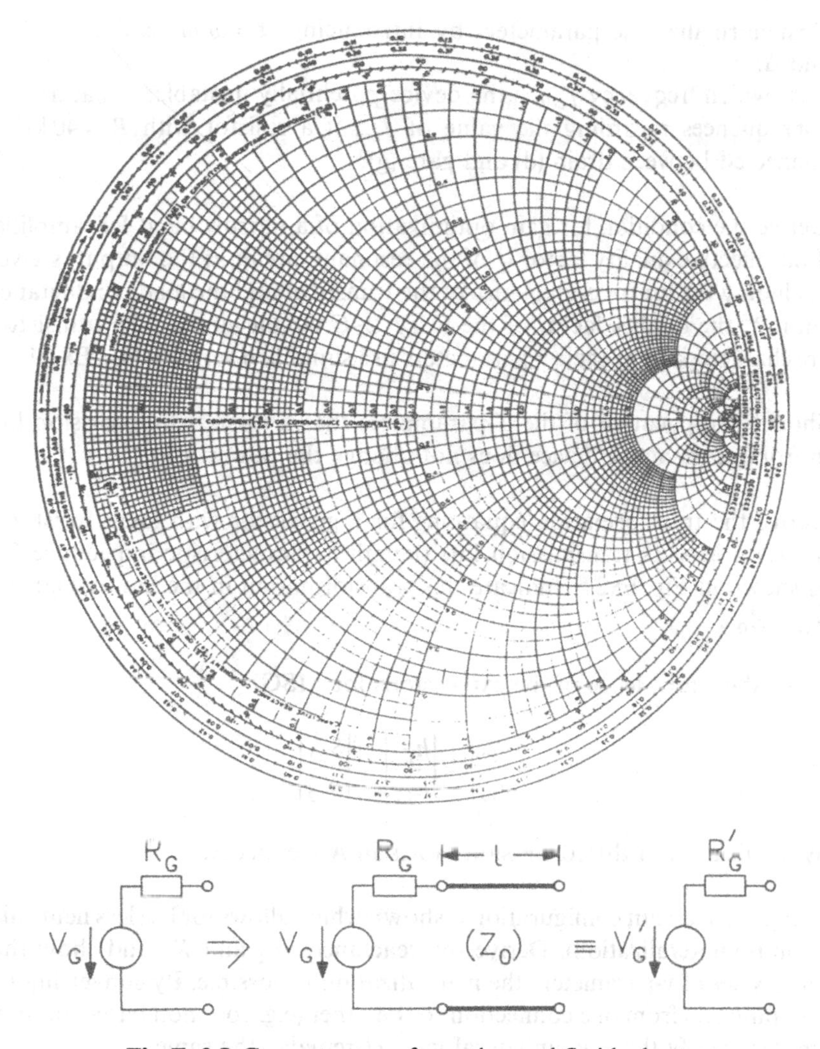

Fig. Ex2.5 Generator transformation and Smith chart.

FET with the complex parameters ($\omega_0 = 2\pi \times 200$ MHz)

$$y_{11} = 100\,\mu\text{S} + \text{j}\omega\frac{1}{\pi}\,\text{pF}$$

$$y_{12} = -\,\text{j}\omega\frac{1}{2\pi}\,\text{pF}$$

$$y_{21} = 16\,\text{mS}\left(1 + \text{j}\frac{\omega}{\omega_0}\right)^{-1}$$

$$y_{22} = 2\,\text{mS}$$

First normalize the parameters by introducing $\Omega = \omega/\omega_0$ and evaluate δ and Δ.

At which frequency f_{crit} is the device potentially unstable? What are the consequences regarding the value of f_{crit} if a resistor with $R = 40\,\mathrm{k\Omega}$ is connected between drain (d) and gate (g)?

2.4 Derive the condition for gain optimization of a narrow band RF amplifier if the effective quality factor Q of the resonant circuit at the output is given (including the inner conductance of the transistor and the load conductance, with Q_0 the quality factor of the lossy L–C circuit alone). The transistors are characterized by their y-parameters in a simplified version (Fig. Ex2.4).

2.5 Show by application of the T-parameters that a lossless transmission line in front of an active device does not change the power gain.

2.6 Derive the transformation equations for a generator with inner resistance R_G via a lossless transmission line (length l, characteristic impedance Z_0) as shown in Fig. Ex2.5. Indicate this transformation in the Smith chart for $R_G = Z_0/2$.

2.7 Show the equivalence of the expressions for MSG

$$\left|\frac{y_{21}}{y_{12}}\right| = \left|\frac{h_{21}}{h_{12}}\right| = \left|\frac{S_{21}}{S_{12}}\right|$$

by application of the conversion tables in Appendix A.

2.8 In Fig. 2.3 a circuit configuration is shown which allows for lossless neutralization (unilateralization). Derive the reactances X_1 and X_2 and show that for any set of y-parameters the neutralization is possible. By converting the y-parameters from one connection to the other (e.g. common base–common emitter), verify that the unilateral gain U remains the same.

3
Noise considerations

In this chapter the circuits aspects of noise are dealt with under the assumption of given noisy amplifier stages. Applying the formalism to derive the implementation of these stages in a transmission chain can be treated without referring to the inner noise sources of the electronic devices involved. These real sources of noise are not considered here.

In Appendix F the corresponding physical noise sources are indicated and a short description of them is given.

A comprehensive collection of papers on noise properties of active devices, characterization and low noise amplifier design can be found in a collection of related papers by Fukui (1981).

3.1 DEFINITIONS

Commonly the noise behaviour of an active device or of an amplifier is characterized by its noise factor F. This quantity is the ratio of the noise power P_{n2} delivered to the load to that noise power P_{n20} which consists of the amplified thermal generator noise alone. Referring to Fig. 3.1(b), where the common amplifier circuit, Fig. 3.1(a), is completed regarding the noise sources, it follows that

$$P_{n20} = G_{eff} P_{n1} \tag{3.1}$$

and

$$F = \frac{P_{n2}}{P_{n20}} = \frac{P_{n2}}{G_{eff} P_{n1}} \tag{3.2}$$

Because

$$G_{eff} P_1 = G_T P_{AVG} \tag{3.3}$$

(section 2.1), also

$$F = \frac{P_{n2}}{G_T P_{nAVG}} = \frac{P_{n2}}{G_T kTB} \tag{3.4}$$

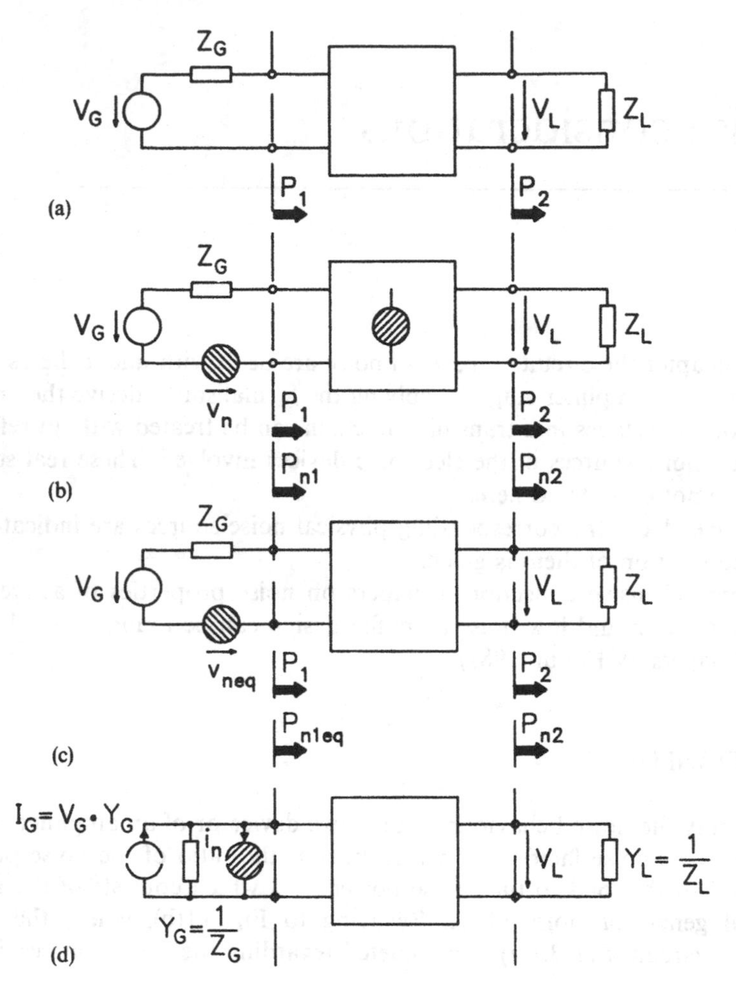

Fig. 3.1 Incorporation of noise sources. (a) Amplifier without noise sources; (b) amplifier with noise sources; (c) amplifier with noise sources transformed to the input; (d) admittance formulation.

can be written, where

$$kTB = P_{\mathrm{nAVG}} = \frac{v_{\mathrm{n}}^2}{4R_{\mathrm{g}}} \tag{3.5}$$

is the available thermal noise power of the generator (k, Boltzmann constant; T, absolute temperature; B, power bandwidth). In this equation, which was derived by Nyquist after measurements of Johnson (Appendix F), v_{n} is the effective magnitude of the thermal noise voltage,

$$v_n = (4kTBR_G)^{1/2} \tag{3.6}$$

where R_G is the real part of the generator impedance $Z_G = R_G + jX_G$.

Instead of a voltage generator a current generator can be considered (Fig. 3.1(d)). With $Y_G = G_G + jB_G$ it follows for the effective value of the noise current

$$i_n = (4kTBG_G)^{1/2} \tag{3.7}$$

and P_{nAVG} can be expressed equivalently as

$$P_{nAVG} = \frac{i_n^2}{4G_G} = kTB \tag{3.8}$$

Noise is a stationary physical process; v_n and i_n are independent from the moment of observation. Therefore the definition

$$v_n^2 = \lim_{\Delta t \to \infty} \frac{1}{\Delta t} \int_{t_0}^{t_0 + \Delta t} \{v_n(t)\}^2 \mathrm{d}t \tag{3.9}$$

is independent of t_0.

The magnitudes of these noise sources v_n, i_n are temperature dependent. At room temperature ($T = 300\,\mathrm{K}$)

$$\frac{v_n}{\mu\mathrm{V}} = 4.07\left(\frac{R_G}{\mathrm{k}\Omega}\right)^{1/2}\left(\frac{B}{\mathrm{MHz}}\right)^{1/2} \tag{3.10}$$

$$\frac{i_n}{\mathrm{pA}} = 4.07\left(\frac{G_G}{\mu\mathrm{S}}\right)^{1/2}\left(\frac{B}{\mathrm{kHz}}\right)^{1/2} \tag{3.11}$$

The term kT in the above equation is valid up to THz frequencies.

More correctly, instead of kT, the quantity

$$E_T = \frac{hf}{e^{hf/kT} - 1} \tag{3.12}$$

should have been used (h, Planck's constant; f, frequency) (Nyquist, 1928). For $hf \ll kT$,

$$E_T \approx kT\left(1 - \frac{hf}{2kT}\right) \tag{3.13}$$

leading to a 0.7% correction at $f = 30\,\mathrm{GHz}$.

Splitting the total noise power P_{n2} into two parts,

$$P_{n2} = P_{n20} + P_{ne} \tag{3.14}$$

one transferred from the generator (P_{n20}), and the other coming from the amplifier (P_{ne}), the excess noise power P_{ne} can be introduced in the formulation. The noise factor

$$F = \frac{P_{n2}}{P_{n20}} \tag{3.15}$$

then becomes

$$F = 1 + F_e \tag{3.16}$$

where

$$F_e = \frac{P_{ne}}{G_{eff}P_{n1}} = \frac{P_{ne}}{G_T P_{nAVG}} = \frac{P_{ne}}{G_T kTB} \tag{3.17}$$

is the excess noise factor.

According to Fig. 3.1(c) the noise contribution of the amplifier can be combined with the generator thermal noise source, as the integral noise power P_{n2} at the output is considered. Because

$$P_{n2} = F(G_T P_{nAVG}) \tag{3.18}$$

this is equal to

$$P_{n2} = G_T\{F P_{nAVG}\} \tag{3.19}$$

which is the output noise power of a noiseless amplifier connected to a generator with extended noise, FP_{nAVG} instead of P_{nAVG}. Interpreting this combined noise as extended thermal noise, available noise power $FkTB$ instead of kTB, it follows for the corresponding equivalent noise voltage that

$$v_{neq}^2 = F4kTR_G B \tag{3.20}$$

In this equation $FT = T_{eff}$ can be written, leading to

$$v_{neq} = (4kT_{eff}R_G B)^{1/2} \tag{3.21}$$

and therefore an effective noise temperature T_{eff} of the system can be introduced. Also with

$$v_{neq} = (4kTR_{eq}B)^{1/2} \tag{3.22}$$

an equivalent noise resistance $R_{eq} = FR_G$ can be defined, or $G_{eq} = FG_G$ in the admittance notation regarding the noise current source in the equation

$$i_{neq} = (4kTG_{eq}B)^{1/2} \tag{3.23}$$

The equivalent input noise power is correspondingly

$$P_{nleq} = FP_{n1} \tag{3.24}$$

A signal power P of the same magnitude would be more or less the detectable limit. Therefore the noise equivalent power

$$\text{NEP} = P_{nleq} \tag{3.25}$$

is a measure of the sensitivity of an amplifier.

The transducer gain G_T is dependent on the generator impedance Z_G, therefore also

$$F = \frac{P_{n2}}{G_T P_{nAVG}} \tag{3.26}$$

becomes dependent on Z_G, and an optimum value $F_{min} = F(Z_{gopt})$ is to expect. To clarify this the noise fourpole after Rothe and Dahlke (1956) will be considered.

3.2 THE NOISE FOURPOLE

The noise generated in a device depends on the inner physical mechanisms involved. Therefore a noise equivalent circuit can be derived containing the physical noise sources in connection with the RF equivalent circuit. This enables us to derive the dependence of the noise behaviour on the biasing condition, etc. Here the noise equivalent circuit after Rothe and Dahlke (1956) is treated which allows one not only to determine experimentally the noise parameters quite easily but also to discuss the overall noise behaviour under a circuit point of view.

If an amplifier stage contains several noise sources then a different noise power P_{n2} may appear at the output if the entrance port is short-circuited or open.

If these noise sources are transformed to the input corresponding to Fig. 3.3(a),

- a noise current source, i_{no}, represents the noise power, P_{n2o} with input open;
- a noise voltage source, v_{ns}, represents the noise power P_{n2s} achieved with the entrance port short-circuited (Fig. 3.2).

These quantities are effective values of the time-dependent noise currents and noise voltages, respectively, as defined earlier.

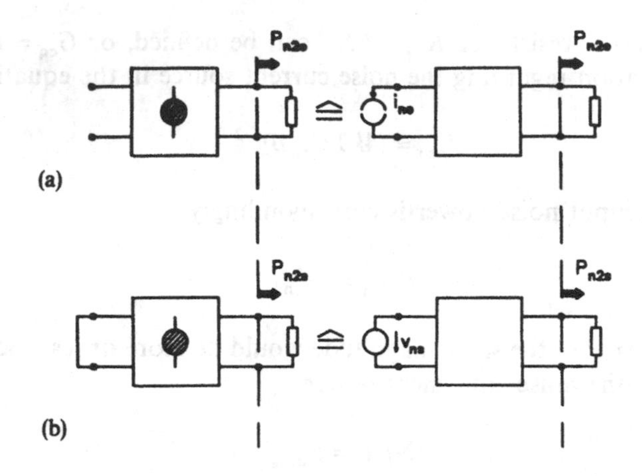

Fig. 3.2 Equivalent noise source at the input of a noisy fourpole. (a) Open input; (b) short-circuited input.

Both sources might be correlated, because a given noise source in the noise fourpole can contribute to P_{2no} as well as to P_{2ns}. This can be proved by determining the quantity

$$\overline{I_{no}(t)V^*_{ns}(t)} \tag{3.27}$$

if $I_{no}(t)$ and $V_{ns}(t)$ represent the time-dependent noise signals.

In general the two noise sources i_{no} and v_{ns} will be correlated, leading to

$$\overline{I_{no}(t)V^*_{ns}(t)} \neq 0 \tag{3.28}$$

Splitting the current source into a noncorrelated part I_{no1} and a fully correlated part I_{no2},

$$I_{no} = I_{no1} + I_{no2} \tag{3.29}$$

it follows that

$$\overline{I_{no}V^*_{ns}} = \overline{I_{no1}V^*_{ns}} + \overline{I_{no2}V^*_{ns}} = \overline{I_{no2}V^*_{ns}} \tag{3.30}$$

Because I_{no2} is fully correlated with the voltage source, it can be written that

$$I_{no2} = Y_C V_{ns} \tag{3.31}$$

which defines the correlation admittance

$$Y_C = G_C + jB_C \tag{3.32}$$

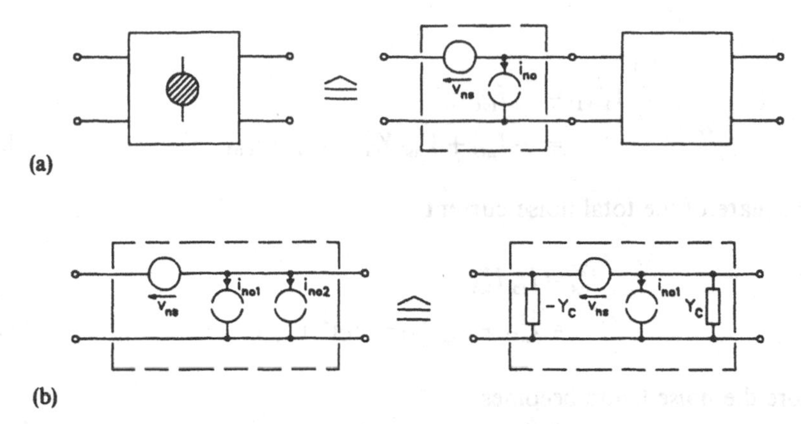

Fig. 3.3 Noisy fourpole. (a) Amplifier with noise sources; (b) equivalent configurations for the noisy fourpole.

The conclusion which can be drawn is, therefore, that four independent quantities describe the noise behaviour of the amplifier, i_{no1}, v_{ns}, G_C and B_C. After Rothe and Dahlke these quantities can be combined in the noise fourpole in front of the (then noiseless) amplifier stage, as indicated in Fig. 3.3. The admittance $-Y_C$ has to be implemented to assure the correct behaviour of the whole circuit corresponding to the given equations. A simple circuit analysis shows that indeed the noise sources i_{no} and v_{ns} do occur at the input, if the latter is open or short-circuited, respectively, corresponding to Fig. 3.2.

This noise fourpole acts together with the generator noise source, leading to i_{tot} instead of i_{nG} alone (Fig. 3.4). To evaluate this combined noise source, the noise current along the short-circuit path 1–1′ has to be determined. It follows

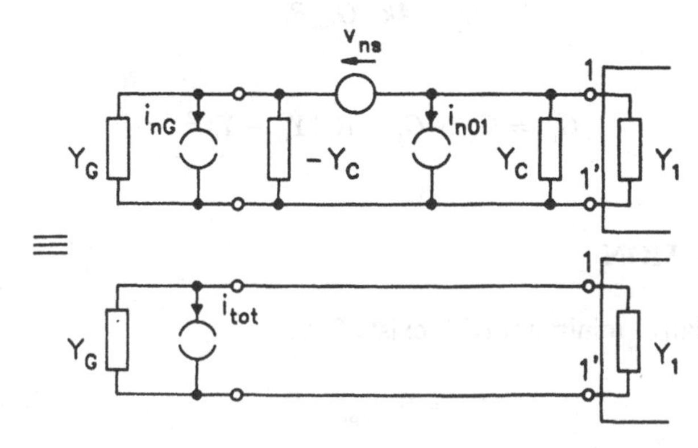

Fig. 3.4 Input noise equivalent circuit including generator and amplifier.

with

$$\begin{aligned} I_{1\to 1'} &= -I_{\text{tot}} \\ &= -I_{\text{nG}} + V_{\text{ns}}(Y_G - Y_C) - I_{\text{nol}} \end{aligned} \tag{3.33}$$

for the square of the total noise current

$$\begin{aligned} i_{\text{tot}}^2 &= I_{\text{tot}} I_{\text{tot}}^* \\ &= i_{\text{nG}}^2 + v_{\text{ns}}^2 |Y_G - Y_C|^2 + i_{\text{nol}}^2 \end{aligned} \tag{3.34}$$

Therefore the noise figure becomes

$$F = \frac{i_{\text{tot}}^2}{i_{\text{nG}}^2} = 1 + \frac{v_{\text{ns}}^2 |Y_G - Y_C|^2 + i_{\text{nol}}^2}{i_{\text{nG}}^2} \tag{3.35}$$

This is equivalent to

$$F = 1 + \frac{R_n |Y_G - Y_C|^2 + G_n}{G_G} \tag{3.36}$$

if the following settings are adopted:

$$\begin{aligned} i_{\text{nG}}^2 &= 4kTG_G B \\ v_{\text{ns}}^2 &= 4kTR_n B \\ i_{\text{nol}}^2 &= 4kTG_n B \end{aligned} \tag{3.37}$$

Correspondingly it can be written that

$$i_{\text{tot}}^2 = 4kTG_{\text{tot}} B, \tag{3.38}$$

leading to

$$G_{\text{tot}} = G_G + G_n + R_n |Y_G - Y_C|^2 \tag{3.39}$$

3.3 OPTIMIZATION

Obviously a relative minimum of F exists for

$$B_G = B_{\text{Gnopt}} = B_C \tag{3.40}$$

leading to

$$F_{\mathrm{Bopt}} = 1 + \frac{G_{\mathrm{n}} + R_{\mathrm{n}}(G_{\mathrm{G}} - G_{\mathrm{C}})^2}{G_{\mathrm{G}}} \tag{3.41}$$

(noise adaptation).

The generator conductance $G_{\mathrm{G}} = G_{\mathrm{Gnopt}}$ best suited to achieve a minimum noise figure follows from

$$\frac{\partial F_{\mathrm{Bopt}}}{\partial G_{\mathrm{G}}} = 0 \tag{3.42}$$

leading to

$$G_{\mathrm{Gnopt}} = \left(G_{\mathrm{C}}^2 + \frac{G_{\mathrm{n}}}{R_{\mathrm{n}}} \right)^{1/2} \tag{3.43}$$

These values G_{Gnopt}, B_{Gnopt} for noise matching differ from G_{Gopt}, B_{Gopt} for maximum power amplification, leading to a lower G_{T} value as for matched input; the **associated gain**

$$G_{\mathrm{optn}} = G_{\mathrm{ass}} \tag{3.44}$$

is smaller than $G_{\max} = \mathrm{MAG}$.

Inserting the optimum values G_{Gnopt}, B_{Gnopt} into the expression for F, the minimum noise figure becomes

$$F_{\min} = 1 + 2R_n \left\{ \left(G_{\mathrm{C}}^2 + \frac{G_{\mathrm{n}}}{R_{\mathrm{n}}} \right)^{1/2} - G_{\mathrm{C}} \right\} \tag{3.45}$$

If we write

$$F = F_{\min} + F_{\mathrm{e}} \tag{3.46}$$

then the excess noise figure F_{e} becomes

$$F_{\mathrm{e}} = \frac{R_{\mathrm{n}}}{G_{\mathrm{G}}} |Y_{\mathrm{nopt}} - Y_{\mathrm{G}}|^2 \tag{3.47}$$

These relationships allow us quite easily to determine the four independent quantities $F_{\min}$, R_{n}, G_{Gnopt}, B_{Gnopt} (e.g. Anastassiou and Strutt, 1974; Bächtold and Strutt, 1967; Fukui, 1966; Rothe and Dahlke, 1956).

Because F represents a power ratio it is commonly expressed in decibels,

$$\frac{F}{\mathrm{dB}} = 10 \log\{F\}$$

The analysis can also be performed by applying the wave concept (Bächtold and Strutt, 1967; Bauer and Rothe, 1956; Fischer and Pfeiler, 1965; Hecken, 1981). The resulting equations for r_{Gnopt}, corresponding to Y_{Gnopt}, and $F = F_{\min} + F_{\mathrm{e}}$ are ($Y_{\mathrm{Gnopt}}Z_0 = y_{\mathrm{Gnopt}} = g_{\mathrm{Gnopt}} + \mathrm{j}b_{\mathrm{Gnopt}}$)

$$r_{\mathrm{Gnopt}} = \frac{1 - g_{\mathrm{Gnopt}}^2 - b_{\mathrm{Gnept}}^2}{(1 + g_{\mathrm{Gnopt}})^2 + b_{\mathrm{Gnopt}}^2} - \mathrm{j}\frac{2b_{\mathrm{Gnopt}}}{(1 + g_{\mathrm{Gnopt}})^2 + b_{\mathrm{Gnopt}}^2} \qquad (3.48)$$

and

$$F = F_{\min} + \frac{Q_{\mathrm{n}}}{1 - |r_{\mathrm{G}}|^2}|r_{\mathrm{G}} - r_{\mathrm{Gnopt}}|^2 \qquad (3.49)$$

where

$$Q_{\mathrm{n}} = \frac{R_{\mathrm{n}}}{Z_0}|1 + y_{\mathrm{Gnopt}}|^2 \qquad (3.50)$$

3.4 GRAPHICAL REPRESENTATION

As well as the available power gain, (section 2.3), the associated noise figure can also be shown in the Smith chart. From

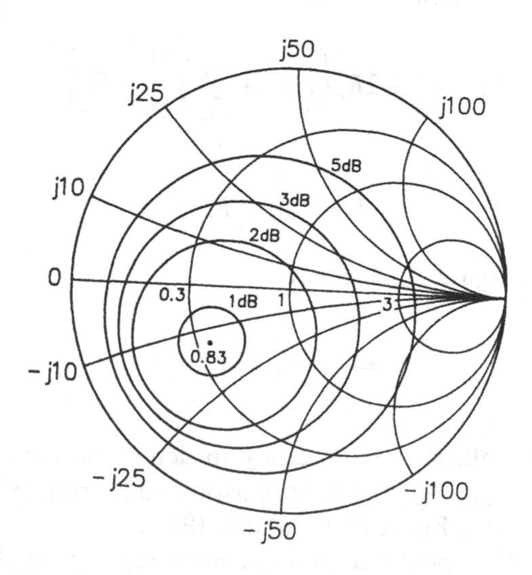

($f = 12\,\mathrm{GHz}$; $Z_{\mathrm{Gnopt}}/Z_0 = 0.449 < -145°$; $V_{\mathrm{DS}} = 2\,\mathrm{V}$; $I_{\mathrm{D}} = 10\,\mathrm{mA}$; associated gain 12.5 dB.)

Fig. 3.5 Noise figure circles of 0.5 μm/200 μm MOCVD grown MODFET. (After Tanaka *et al.*, 1986.)

$$F = F_{\min} + \frac{R_n}{G_G} |Y_{\text{Gnopt}} - Y_G|^2 \tag{3.51}$$

it becomes clear that in the rectangular $\{G_G, B_G\}$ plane the curves $F = \text{const.}$ are circles around the optimum generator admittance $Y_{\text{Gnopt}} = G_{\text{Gnopt}} + jB_{\text{Gnopt}}$ analogous to the power amplification (Rieke diagram). Their transformation into the r_G plane,

$$r_G = -\frac{y_G - 1}{y_G + 1}, \quad y_G = Y_G Z_0 \tag{3.52, 3.53}$$

leads again to circles, as shown for an example in Fig. 3.5. Their coordinates can be expressed either by the admittances or S-parameters as in case of the G_{AV} circles (section 2.3).

Identical equations for r, φ_p are achieved by setting Y_{nopt} instead of Y_{opt} and

$$\delta_{Gn} = \frac{|y_{21}^2|}{2g_{22}} Z_0 (F - F_{\min}) \tag{3.54}$$

instead of δ_G.

3.5 NOISE FIGURE AND NOISE MEASURE

If cascaded amplifier stages are used, the integral noise figure $F = F_\Sigma$ is of interest. As shown first by Friis (1944), an amplifier chain, (Fig. 3.6(a)), exhibits a noise

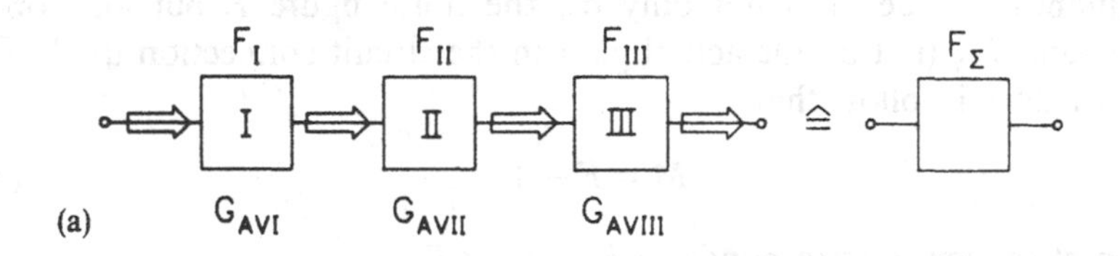

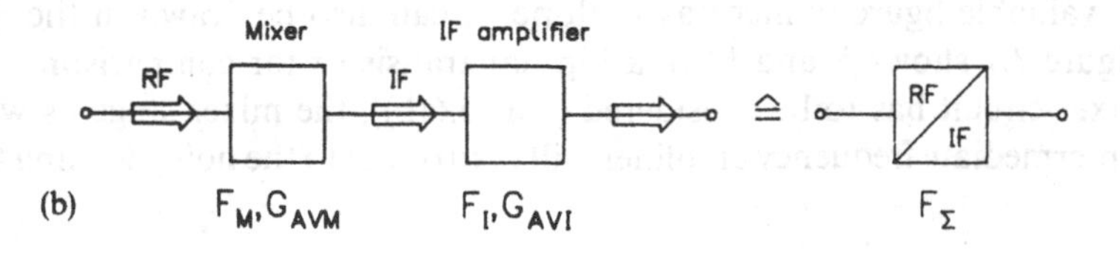

Fig. 3.6 Cascaded amplifier. (a) Linear chain; (b) mixer.

figure of

$$F_\Sigma = F_\mathrm{I} + \frac{F_\mathrm{II} - 1}{G_\mathrm{AVI}} + \frac{F_\mathrm{III} - 1}{G_\mathrm{AVI} G_\mathrm{AVII}} + \cdots \tag{3.55}$$

(the noise figure of the following stage has to be measured/computed with an input generator impedance equivalent to the output impedance of the foregoing stage). This quantity is dependent on the sequence of the amplifier stages, as in case of a two-stage system

$$F_\mathrm{I,II} = F_\mathrm{I} + \frac{F_\mathrm{II} - 1}{G_\mathrm{AVI}} \tag{3.56}$$

will be in general different from

$$F_\mathrm{II,I} = F_\mathrm{II} + \frac{F_\mathrm{I} - 1}{G_\mathrm{AVII}} \tag{3.57}$$

As pointed out by Haus and Adler (1959), the sequence leading to the lowest integral noise figure F_Σ is a series connection I, II, III, corresponding to

$$M_\mathrm{I} < M_\mathrm{II} < M_\mathrm{III} \cdots \tag{3.58}$$

where M is the noise measure

$$M = \frac{F - 1}{1 - (1/G_\mathrm{AV})} \tag{3.59}$$

of the different stages. Therefore the amplifier stage exhibiting the lowest M-value has to be used as the front end.

This quantity M depends not only on the noise figure F but also on the available gain G_AV (not on the actual gain in the circuit connection used). Only for $G_\mathrm{AV} \gg 1$ does it follow that

$$M = F - 1 \tag{3.60}$$

and the best sequence corresponds to $F_\mathrm{I} < F_\mathrm{II} < F_\mathrm{III} \cdots$.

Therefore this quantity M, which contains the available gain of the device, is the more valuable figure of merit as F alone. It can also be drawn in the Smith chart. Figure 3.7 shows F and M of a bipolar transistor for comparison.

If a mixer circuit has to be considered (Fig. 3.6(b)), the mixer stage as well as the first intermediate frequency amplifier will contribute to the noise, leading to

$$F_\Sigma = F_{\Sigma\mathrm{AM}} = F_\mathrm{M} + \frac{F_\mathrm{I} - 1}{G_\mathrm{AVM}} \tag{3.61}$$

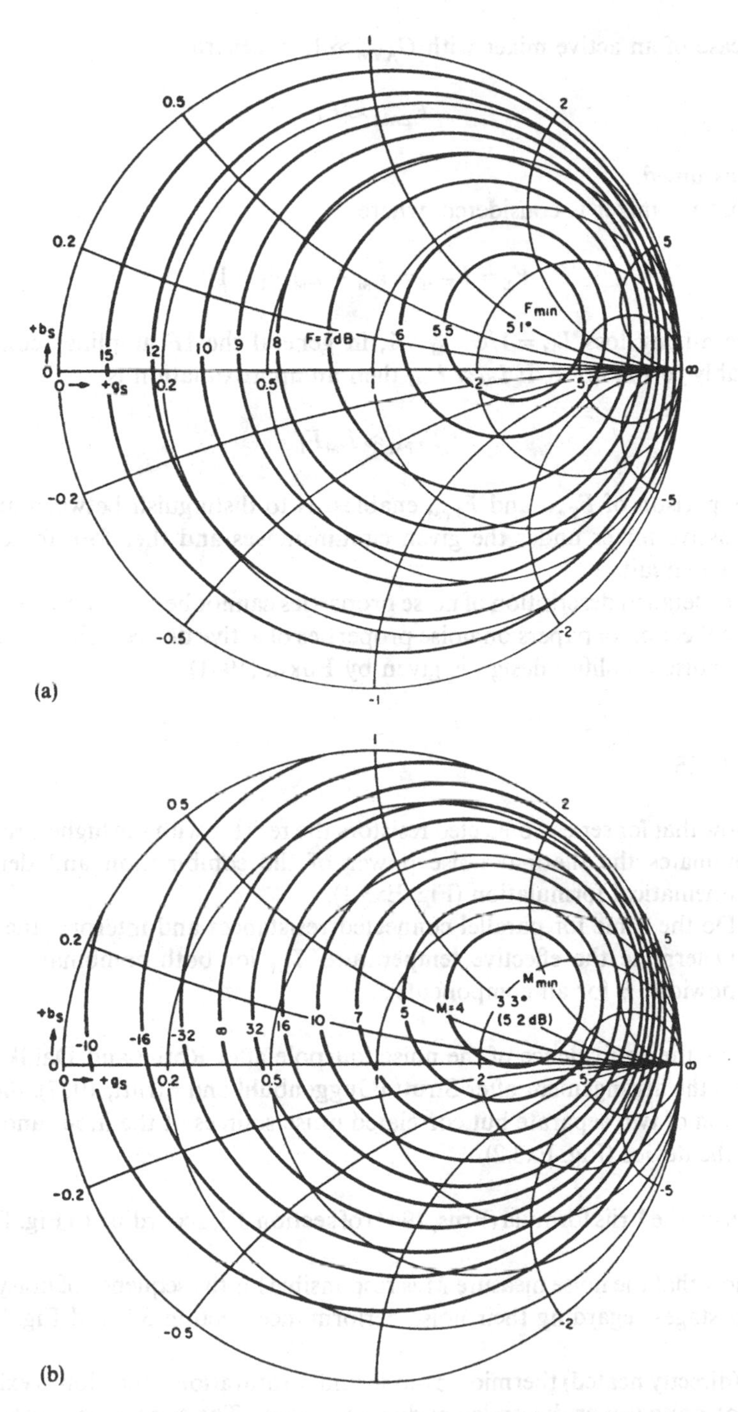

Fig. 3.7 Noise representation of a silicon n–p–n transistor ($f = 1.3$ GHz) in the complex r_g plane. (a) Noise figure F; (b) noise measure M. After Bächtold and Strutt, 1967.)

In the case of an active mixer with $G_{AVM} \gg 1$, in general

$$F_{\Sigma AM} \approx F_M \tag{3.62}$$

can be assumed.

If a passive mixer is considered, where

$$F_\Sigma = F_{\Sigma PM} = F_M + L_M(F_I - 1) \tag{3.63}$$

with the mixing loss $L_M = 1/G_{AVM} \gg 1$, in general the IF amplifier contributes remarkably to the noise. If $L_M \approx F_M$, then an approximation is

$$F_{\Sigma PM} \approx L_M F_I \tag{3.64}$$

The comparison of $F_{\Sigma AM}$ and $F_{\Sigma PM}$ enables us to distinguish between an active and a passive mixer under the given circumstances and therefore to select the best suited circuit.

A more detailed description of noise properties cannot be given here. A comprehensive collection of papers on noise properties of active devices, characterization and low noise amplifier design is given by Fukui (1981).

EXERCISES

3.1 Show that for series-connected resistors the resistor with the highest resistance dominates the thermal noise power of the combination and derive the mathematical formulation (Fig. Ex3.1).

Do the same for parallel connected resistances and interpret the result.

Determine the effective temperature T_{eff} for both combinations (same bandwidth B for all components).

3.2 Show the equivalence of the noise fourpole after Rothe and Dahlke (1956) with the arrangement after Strutt (Guggenbühl and Strutt, 1957), the application of two separate but correlated noise sources at the input and output of the device (Fig. Ex3.2).

3.3 Derive the Friis formula (Friis, 1944) of section 3.5, according to Fig. Ex3.3.

3.4 Show that the noise measure M is responsible for the sequence of noisy amplifier stages regarding their noise performance (section 3.5 and Fig. Ex3.3).

3.5 A (directly heated) thermionic diode under saturation condition is exhibiting shot noise (inner dynamic conductance zero). The tube is connected to an ohmic resistor R_G (Fig. Ex3.4).

Show that this combination acts as a noise source with

$$(F+1)P_{n0} = F^*P_{n0}$$

where P_{n0} is the thermal Johnson noise of the resistance.

Determine the value of R_G in ohms, if $F \approx (1 + I_D/\text{mA})$ with I_D the DC current through the diode. Evaluate the effective noise temperature T_{eff} of the resulting noise generator. Determine the frequency limit, given by the RC time constant, if $C_D = 10\,\text{pF}$ is the parasitic capacitance of the diode. Why is the computed value of R_G especially suitable for the connection to a $Z_0 = 50\,\Omega$ transmission line?

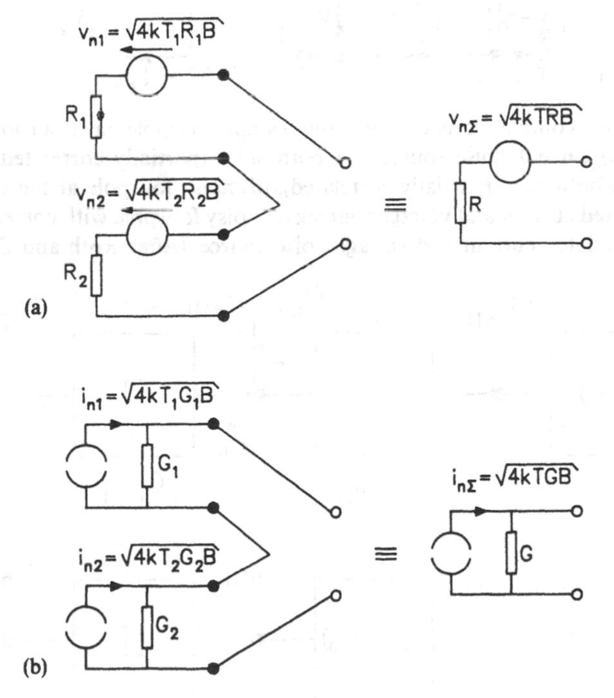

Fig. Ex3.1 Combination of noisy components. (a) Series connection of two resistances R_1, R_2; (b) parallel connection of two conductances G_1, G_2.

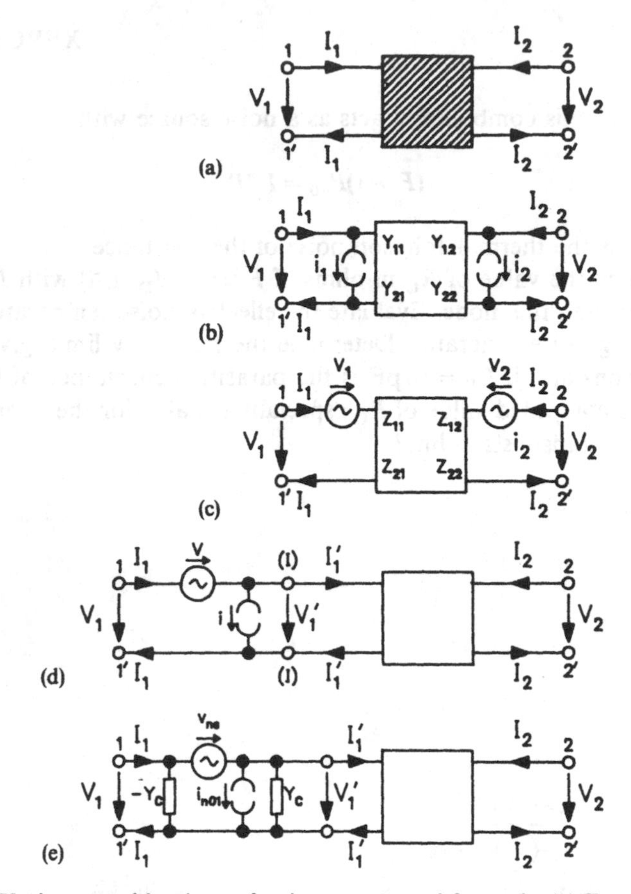

Fig. Ex3.2 Various combinations of noise sources and fourpole. (a) Fourpole with inner noise sources; (b) current noise sources on both sides (partially correlated); (c) voltage noise sources on both sides (partially correlated); (d) noisy fourpole at the entrance port (partially correlated current and voltage source); (e) noisy fourpole with correlation admittance and uncorrelated current and voltage noise source. (After Roth and Dalke, 1956.)

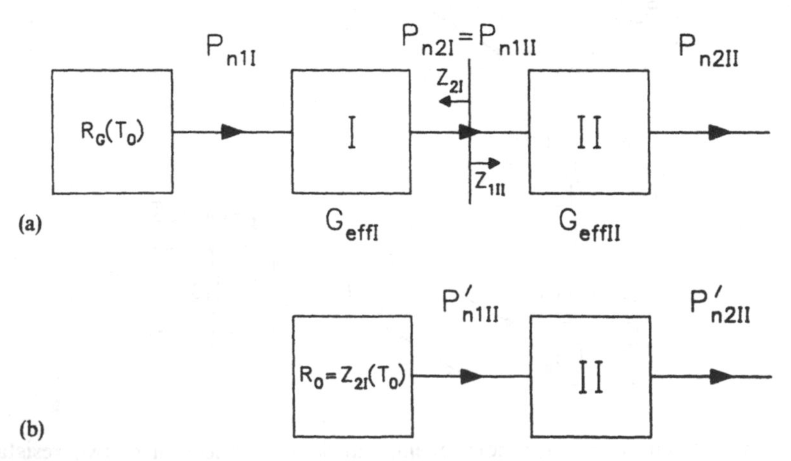

Fig. Ex3.3 Cascaded noisy fourpoles of an amplifier chain. (a) Fourpole connection; (b) measurement procedure for F_{II}.

$R \gg R_G$ C I_D V_{Batt} $R_G(T_0)$ ≡ $R_G((F+1)T_0)$

(a)

C_D $R_G(F^*T_0)$

(b)

Fig. Ex3.4 Noise source with thermionic diode ($1/\omega C \ll R_G$ for the frequency range of interest). (a) Circuit and equivalent noise source; (b) effective admittance ($1 + F = F^*$).

4

Gain and bandwidth

The achievable gain G and bandwidth B of an amplifier depend on the active device itself as well as on the circuit configuration. The latter is especially important if feedback is considered. The feedback principle enables us to modify the original characteristics of an active device implemented in the circuit and will be discussed first.

4.1 FEEDBACK

Dependent on the phase angle of the feedback loop, different behaviour occurs: **negative** feedback if the feedback signal reduces the input signal, or **positive** feedback if the input signal is enhanced.

Negative feedback can smooth the characteristics and therefore linearize the transfer function and extend the 3 dB cutoff frequency f_c (frequency where the power gain drops to 50% of the low frequency value). Furthermore, negative feedback enhances the circuit immunity against any possible change of bias, temperature, ageing, etc. Conversely, positive feedback enhances the non-linear behaviour and makes the circuit more sensitive to possible changes.

Not only does feedback modify the transfer characteristics but also the effective impedances of the amplifier. This is of great importance for small signal processing as in case of operational amplifiers. In Table 4.1 the different types of feedback are indicated and their influence on the input and output impedance is shown. Referring to Appendix A these different types of feedback correspond to different fourpole couplings.

The ideal behaviour shown in Table 4.1, however, is modified in practice by parasitic components and phase shifts at higher frequencies. Then negative feedback turns to positive feedback which needs special attention (section 4.1.2).

According to Fig. 4.1 an ideal feedback system consists of the original amplifier with its transfer function

$$G_o = \frac{Q_{out}}{Q_{in0}} \tag{4.1}$$

and a feedback loop.

Table 4.1 Negative feedback amplifiers

	Transimpedance amplifier	Current amplifier
Type of feedback	Voltage to current feedback	Current to current feedback
Change of input resistance	$R_i \to 0$	$R_i \to 0$
Change of output resistance	$R_o \to 0$	$R_o \to \infty$
Principal circuit configuration	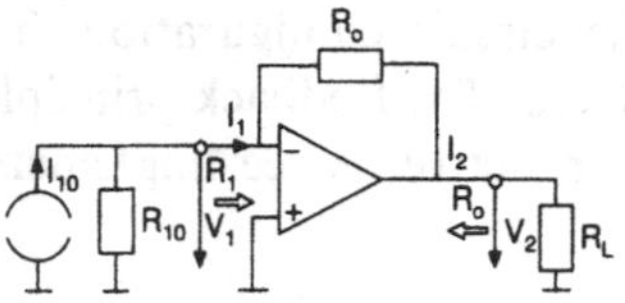	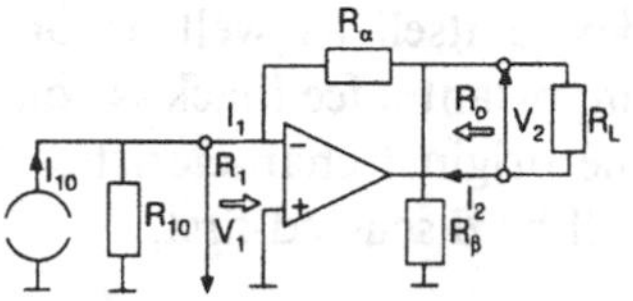
Typical operational amplifier circuit	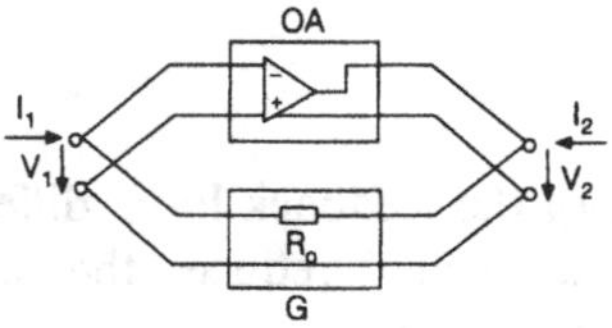	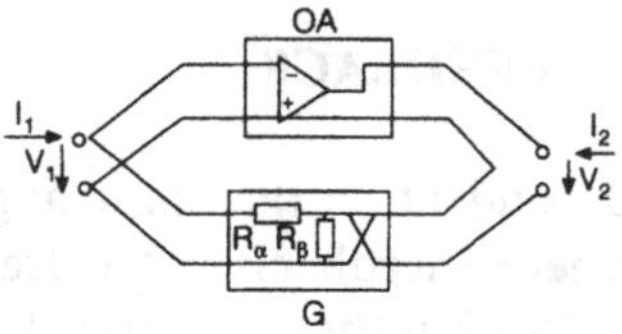
Corresponding fourpole equations	$I_1 = y_{11}V_1 + y_{12}V_2$ $I_2 = y_{21}V_1 + y_{22}V_2$	$I_1 = p_{11}V_1 + p_{12}I_2$ $V_2 = p_{21}V_1 + p_{22}I_2$
Matrix forms	$\boldsymbol{Y} = \boldsymbol{Y}_{op} + \boldsymbol{Y}_G$	$\boldsymbol{P} = \boldsymbol{P}_{op} + \boldsymbol{P}_G$
Transfer function	$\frac{V_2}{I_1} = -R_0$	$\frac{I_2}{I_1} = \frac{R_\alpha + R_\beta}{R_\beta}$
Type of feedback	Voltage to voltage feedback	Current to voltage feedback
Change of input resistance	$R_i \to \infty$	$R_i \to \infty$
Change of output resistance	$R_o \to 0$	$R_o \to \infty$
Principal circuit configuration	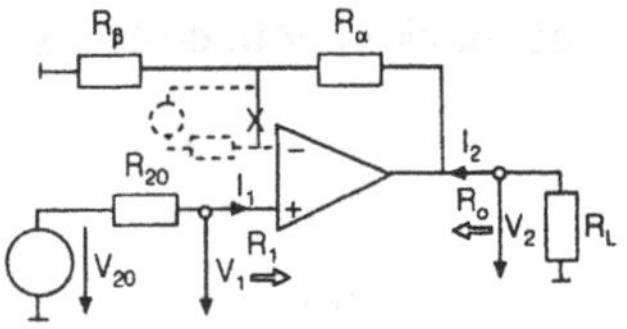	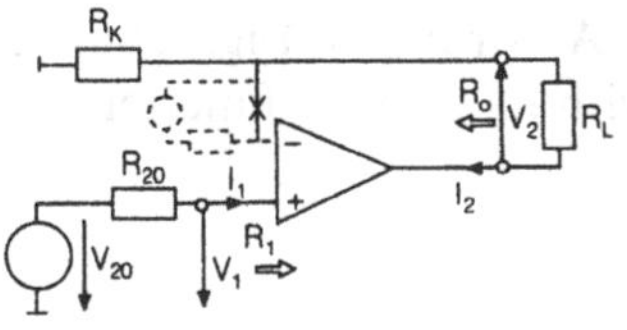

Table 4.1 (*Contd.*)

	Transimpedance amplifier	Current amplifier
Typical operational amplifier circuit	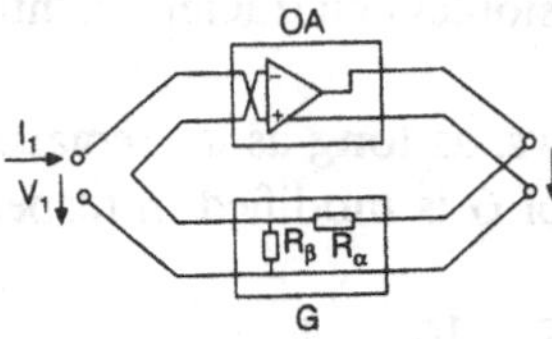	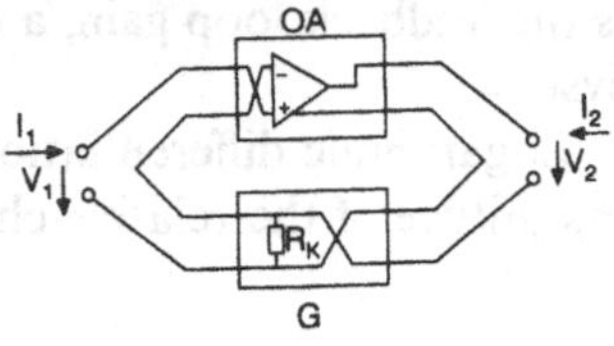
Corresponding fourpole equations	$V_1 = h_{11}I_1 + h_{12}V_2$ $I_2 = h_{21}I_1 + h_{22}V_2$	$V_1 = z_{11}I_1 + z_{12}I_2$ $V_2 = z_{21}I_1 + z_{22}I_2$
Matrix forms	$\boldsymbol{H} = \boldsymbol{H}_{op} + \boldsymbol{H}_G$	$\boldsymbol{Z} = \boldsymbol{Z}_{op} + \boldsymbol{Z}_G$
Transfer function	$\frac{V_2}{V_1} = \frac{R_\alpha + R_\beta}{R_\beta}$	$\frac{I_2}{V_1} = -\frac{1}{R_k}$

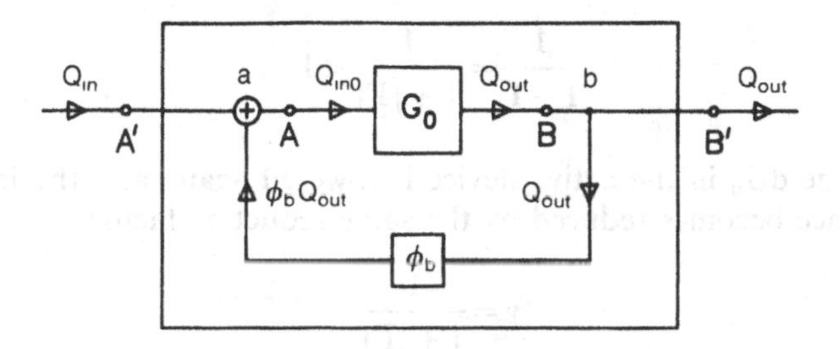

Fig. 4.1 Feedback system.

The quantities Q_{in0}, Q_{out} might be any quantity of interest (pressure, voltage, current, temperature, etc.). Via the output node b the feedback loop is coupled lossless to the output quantity $Q_{out}(B') \approx Q_{out}(B)$ and linearly coupled back via the transformation ϕ_b, added at node a to the input quantity. As a result a modified system, $A' - B'$ instead of $A - B$, is created with some unique properties.

With

$$Q_{in0} = Q_{in} + \phi_b Q_{out} \tag{4.2}$$

it follows for the modified system that

$$G = \frac{Q_{out}}{Q_{in}} = \frac{Q_{out}}{Q_{in0} - \phi_b Q_{out}} = \frac{G_0}{1 - \phi_b G_0} \tag{4.3}$$

The quantity

$$\phi_b G_0 = \Gamma \tag{4.4}$$

is the feedback loop gain, a dimensionless characteristic number of the feedback system.

Logarithmic differentiation shows, as long as ϕ_b remains unaffected, that the magnitude of the relative change of G is modified in respect to that of G_0,

$$\frac{dG}{G} = \frac{dG_0}{G_0} \frac{1}{1-\Gamma} \tag{4.5}$$

whatever the source of the influence on G_0 might be. The same is true for higher order differential coefficients.

To demonstrate the consequences, an electric system containing an active device with power gain $G_0 > 0$, neglecting phase shifts, will be considered. If $0 < \Gamma < 1$, then positive feedback occurs and G becomes enhanced, but also relative changes are enlarged. If on the other hand, $\Gamma < 0$, than negative feedback occurs. The gain G becomes smaller than G_0, because

$$\frac{1}{1-\Gamma} = \frac{1}{1+|\Gamma|} < 1 \tag{4.6}$$

but any change dG_0 in the active device is lowered again, and the influence of any disturbance becomes reduced by the same reduction factor

$$\gamma = \frac{1}{1+|\Gamma|} \tag{4.7}$$

If the magnitude of the feedback loop gain is very large, $|\Gamma| \gg 1$, it follows that

$$G = \frac{G_0}{|\Gamma|} \tag{4.8}$$

or

$$G = \frac{1}{|\phi_b|} \tag{4.9}$$

Therefore the overall gain becomes independent of the original amplifier, which must of course have a high amplification G_0. Because the reduction factor γ then becomes very small, $\gamma \ll 1$, the application of strong negative feedback allows us to linearize transfer functions, to reduce effectively drift effects, etc. An essential condition, however, is the consistency of ϕ_b, independent of the sources of influences modifying the quantity G_0.

The case $\phi_b G_0 = 1$ needs special attention as it leads to unstable behaviour of the system, as shown in the following section. In Appendix D, as an example of unstable behaviour, an oscillator circuit is treated. Linear circuit theory allows us to study the corresponding turn-on behaviour, whereas the final saturation mode, e.g. stable oscillations, can be evaluated by a quasistatic approach (section 6.2).

In the following, the situation in case of negative feedback is discussed, if Γ becomes frequency dependent.

4.1.1 Negative feedback

For simplicity, a circuit will be considered where the amplifier exhibits a given frequency-dependent transfer characteristic, whereas the feedback loop shows no frequency dependence ($\phi_b = \text{const.}$). Taking into account the frequency dependence of the gain $G_0 = G_0(f)$ and therefore including phase shifts as well, the frequency behaviour of the feedback system can be studied. As a result the bandwidth B is enlarged if negative feedback is applied. This can be shown, e.g. in a log–log plot of the magnitude of the power gain in dependence on frequency (Fig. 4.2).

Assuming a onepole configuration, e.g. RC limited, $\Omega = \omega RC$, then

$$G_0 = \frac{G_{00}}{1 + j\Omega} \tag{4.10}$$

The magnitude

$$|G_0| = \frac{G_{00}}{(1 + \Omega^2)^{1/2}} \tag{4.11}$$

becomes

$$|G_0| = G_{00} \quad \text{for} \quad \Omega \ll 1 \tag{4.12}$$

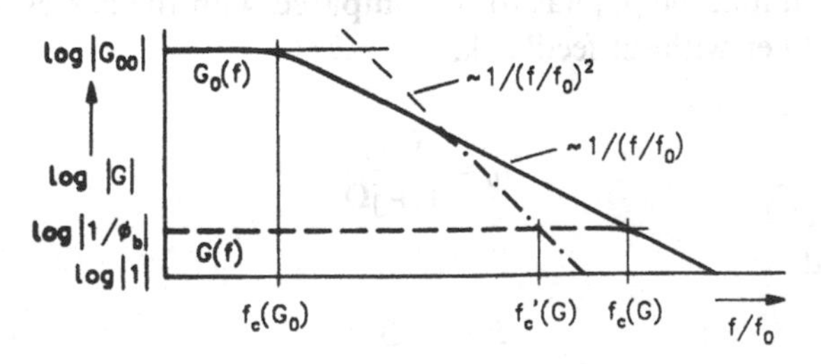

Fig. 4.2 Gain diagram ($f_0 = 1/2\pi RC$). Indicated on the log–log scale are the frequency dependencies of the gain G_0 of an amplifier without feedback (——), with feedback (G, – – –) and with a twopole dependency (–·–·–·).

and

$$|G_0| = \frac{G_{00}}{\Omega} \quad \text{for} \quad \Omega \gg 1 \tag{4.13}$$

Figure 4.2 indicates, on a log–log scale, the frequency dependencies of the gain $|G_0|$ of an amplifier with and without feedback, and with a twopole dependency. The low frequency value $|G| = 1/|\phi_b|$ of the amplifier with large feedback, $|\Gamma| \gg 1$, is much lower than the gain G_{00} of the amplifier without feedback. However, the bandwidth B is correspondingly enlarged,

$$B = f_c(G) \gg B_0 = f_c(G_0) \tag{4.14}$$

where

$$G = \frac{G_0(f)}{1 + |\phi_b| G_0(f)} \tag{4.15}$$

The modified gain is

$$G = \frac{G_{00}}{(1 + j\Omega)\left(1 + |\phi_b| \dfrac{G_{00}}{1 + j\Omega}\right)} \tag{4.16}$$

$$= \frac{G_{00}}{1 + |(\phi_b)| G_{00} + j\Omega} \tag{4.17}$$

Because $|\phi_b| G_{00} = \Gamma_0 \gg 1$ it follows that, for the frequency dependence of the feedback amplifier,

$$G = \frac{1}{|\phi_b|} \frac{1}{1 + j(\Omega/\Gamma_0)} \tag{4.18}$$

The slope of its amplitude $|G|$ has to be compared with the corresponding slope $|G_0|$ of the amplifier without feedback,

$$G_0 = \frac{G_{00}}{1 + j\Omega} \tag{4.19}$$

It follows that

$$B = \Gamma_0 B_0 \tag{4.20}$$

with B the enlarged 3 dB cutoff frequency of the feedback amplifier, Γ_0 being the low frequency value of $|\Gamma|$.

The gain–bandwidth product, however, remains unchanged:

$$BG = B\frac{1}{|\phi_b|} = B_0 G_{00} \tag{4.21}$$

giving the opportunity to exchange gain for bandwidth by negative feedback.

It should be mentioned that the latter relationship is only valid for onepole networks. If more than one time constant becomes involved, again $B|G|$ is larger than without negative feedback; however, $B|G| < B|G_0|$.

This is indicated in Fig. 4.2 for a twopole configuration,

$$G_0 = G_{00}\frac{1}{1+j\Omega}\frac{1}{1+j\Omega_2} \tag{4.22}$$

with

$$f_c' < f_c \tag{4.23}$$

So far no complex values of ϕ_b have been considered. In practice the frequency-dependent phase lag of the gain G_0 results in a phase shift in the closed feedback loop. This change, from

$$\Gamma = -\Gamma_0$$

to

$$\Gamma = -|\Gamma(f)|e^{-j\varphi_\Gamma(f)} \tag{4.24}$$

leads to instabilities as the negative feedback turns to positive feedback at higher frequencies.

Self-oscillation can only be avoided if the magnitude $|\Gamma|$ of the closed feedback loop gain becomes smaller than unity in this critical frequency range. The limit of stability is therefore

$$|\Gamma| = 1 \text{ at } \varphi_\Gamma = \pi \tag{4.25}$$

Only for $|\Gamma| < 1$ at $\varphi_\Gamma = \pi$ is stability achieved. This can be shown graphically in the Bode diagram, a combination of the representation of $|G_0|$ and $|G|$ together with the frequency dependence of the phase angle $\varphi_\Gamma(f)$ as shown in Fig. 4.3.

Because

$$|\Gamma| = |\phi_b||G_0| = \frac{1}{|G|}|G_0| \tag{4.26}$$

the critical magnitude $|\Gamma| = 1$ is given at the crossover point where $|G_0| = |G|$. At this frequency the magnitude of the phase angle φ_Γ must for stable behaviour be smaller than π.

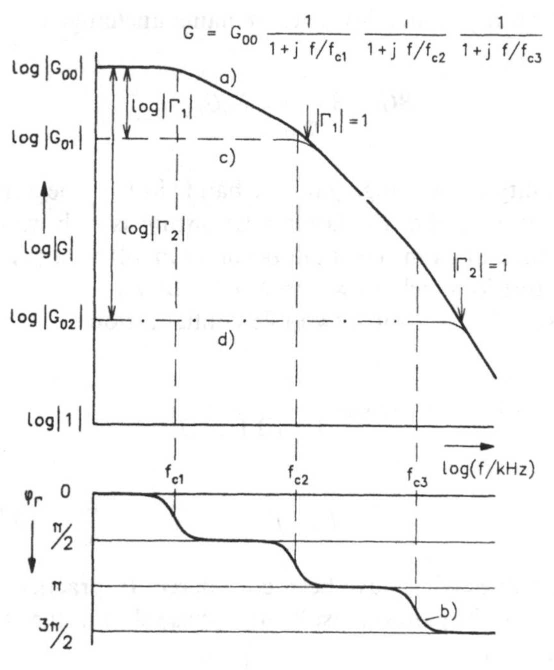

Fig. 4.3 Example of a Bode diagram with three different time constants $\tau_1 > \tau_2 > \tau_3$ governing the frequency behaviour of the amplifier, $f_{c1} = 1/2\pi\tau_1 < f_{c2} = 1/2\pi\tau_2 < f_{c3} = 1/2\pi\tau_3$. (a) Frequency dependence $|G_0(f)|$ of the original amplifier; (b) phase angle of the original gain; each time constant contributes with $(\Delta\varphi_\Gamma)_{max} = \pi/2$; (c) stable configuration: $|\Gamma_1(\varphi_\Gamma = \pi)| < 1$ (——); (d) unstable configuration: $|\Gamma_2(\varphi_\Gamma = \pi)| > 1$ (–·–·–·).

In Fig. 4.3 two different cases are indicated. Curve (c) belongs to a stable situation, as the corresponding value G_{01} crosses the gain curve $|G_0(f)|$ at a phase angle $|\varphi_\Gamma| < \pi$. Contrary to this, at the larger feedback coupling, curve (d), the crossing point shows $|\varphi_\Gamma| > \pi$, corresponding to unstable behaviour. By changing the frequency characteristics of the feedback loop, again stable behaviour may occur, but will not be considered here. These modifications of the original behaviour are of great importance in case of operational amplifiers (e.g. Millmann and Grabel, 1988).

4.1.2 Positive feedback

As mentioned earlier, positive feedback reduces the magnitude of the input signal needed for given output but enhances – besides the amplification – the sensitivity regarding any changes of the original amplifying system. This follows from

$$\frac{dG}{G} = \frac{dG_0}{G_0}\frac{1}{1-\Gamma} > \frac{dG_0}{G_0} \tag{4.27}$$

because $\Gamma > 0$. The critical value of the loop gain $\Gamma = \phi_b G_0 = 1$ marks the border of stable and unstable operation. In the case of an oscillator or the switching in a static memory cell, the turn-on behaviour is governed by the amount of excess loop gain available in the circuit. This can be understood by discussing the differential equation of the linearized system in terms of the complex frequency $p = \sigma + j\omega$. At growing amplitude the nonlinear behaviour of the system reduces the feedback loop gain to unity, leading to the final stable amplitude of oscillations or the 'on' or 'off' state of a memory cell (flip-flop), respectively. These topics are described in Appendix D.

EXERCISES

4.1 Given the circuit of Fig. Ex4.1, which type of feedback is realized, and which transfer quantity is stabilized?

Determine the voltage gain $G'_V = V_2/V'_1$ of the feedback amplifier.

Derive the relative variation of G'_V in relation to $G_V = V_2/V_1$.

Which K-value is necessary ($K = V_3/V_4$) to achieve a tenfold reduction of any disturbing influence? Indicate a technical realization in the case of a low frequency amplifier stage with $G_V = 100$.

4.2 Show that for a system with negative feedback the same reduction factor is valid for all terms of the Taylor series of a nonlinear transfer function (real parameters, (section 4.1)).

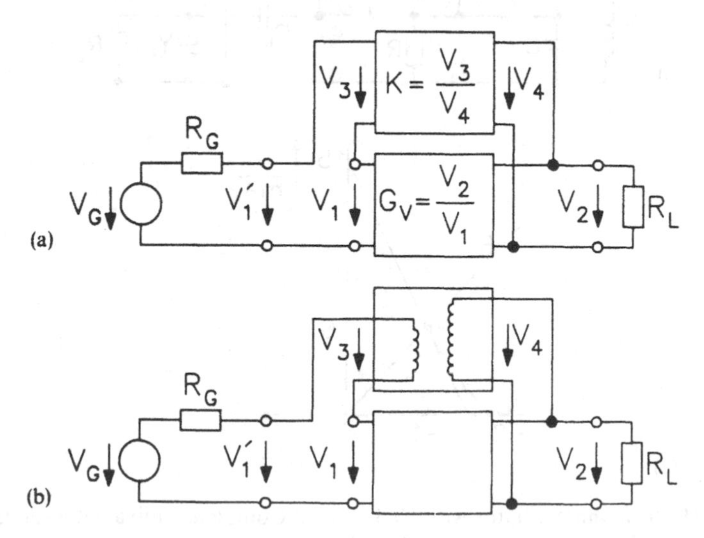

Fig. Ex4.1 Feedback system. (a) Principle of the feedback circuit; (b) Practical version with transformer.

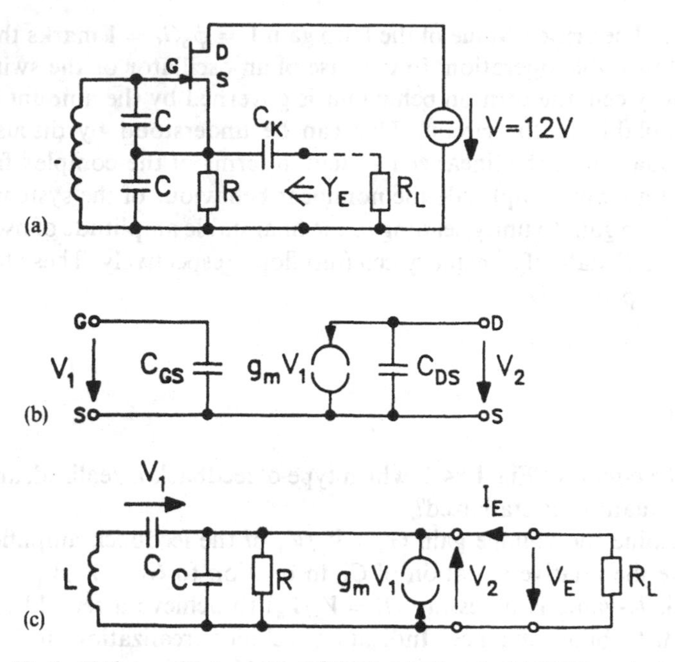

Fig. Ex4.2 Huth–Kühn oscillator with FET (I). (a) Circuit; (b) FET equivalent circuit; (c) Simplified equivalent circuit.

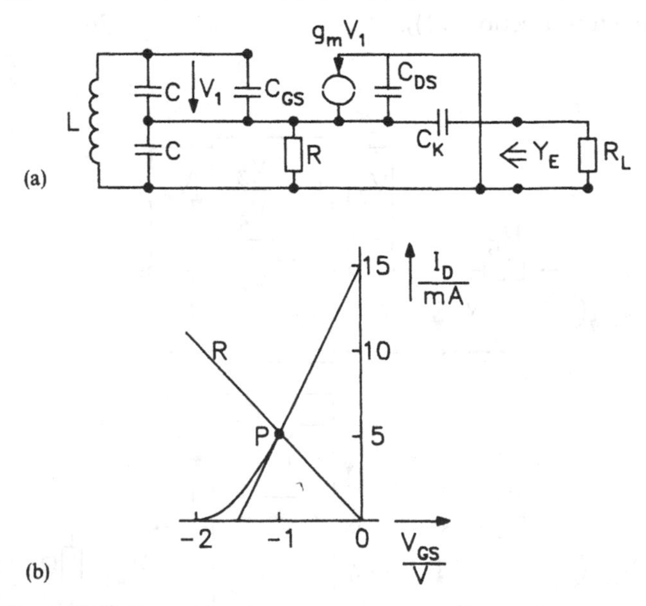

Fig. Ex4.3 Huth–Kühn oscillator with FET (II). (a) Complete equivalent circuit; (b) FET transfer characteristic $I_D = I_D(V_{GS})$ with load line.

4.3 Derive the gain G for an amplifier with three identical stages (onepole transfer functions). Discuss its stability and gain behaviour depending on the applied feedback over the whole amplifier (section 4.1.1). Show that differences exist if the feedback is applied to the three stages individually.

4.4 Analyse a Huth–Kühn oscillator by applying the y-parameters of the active device and the passive feedback fourpole (Figs Ex4.2 and Ex4.3):

First determine the absolute value of R for $I_D = 5\,\text{mA}$ and give the value of the transconductance

$$g_m = \frac{dI_D}{dV_{GS}}$$

at this bias condition (Fig. Ex4.3). Draw the whole equivalent circuit ($C \gg C_{GS}, C_{DS}$; $C_K \to \infty$).

Determine the output admittance Y_E and compute the frequency f_0 where $\text{Im}\{Y_E\} = 0$. Give the resulting real part $\text{Re}\{Y_E\}$ at $f = f_0$ and determine the minimum resistance R_L for onset of oscillations.

4.5 Use the method of coupled fourpoles to analyse a Hartley oscillator, and determine its behaviour regarding the onset of oscillations (also Exercise 4.4).

The circuit is given in Fig. Ex4.4. It can be understood as a series–parallel

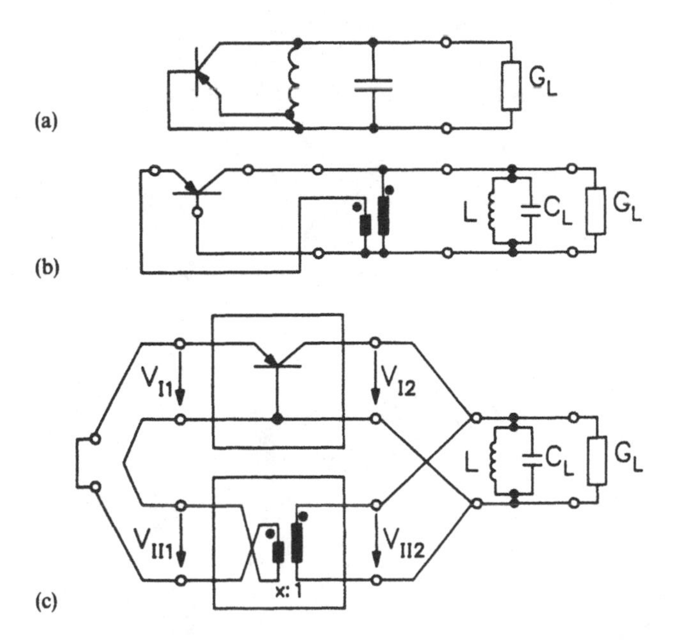

Fig. Ex4.4 Hartley oscillator (RF part). (a) Circuit arrangement; (b) circuit with extracted ideal transformer; (c) equivalent series–parallel connection of two fourpoles.

coupling of an active and passive (feedback) fourpole. Show the resulting combination and derive the **H**-matrix with the parameter $x = -V_{\text{II}1}/V_{\text{II}2}$.

Derive the formula for the resulting output admittance ($Z_G = 0$) and simplify the equation for a bipolar transistor (common base) with the parameters

$$h_{11}, h_{12} = 0, \quad h_{21} = -1, \quad h_{22}$$

Show that an optimum value of x exists and derive the corresponding value of the maximum negative conductance G_N achievable and the occurring negative capacitance C_N.

Express the maximum load conductance $G_{L\max}$ for onset of oscillations by the transistor parameters

$$y_{11} = G_d + j\omega C_d, \quad G_d = \frac{I_E}{V_T}$$

Derive also the capacitance C which determines together with the inductance L the oscillation frequency f_0. What is the value of the growth constant σ for $G_L < G_{L\max}$?

5
Device-related parameters

Besides the general fourpole and twoport parameters (Chapter 1), special device-related parameters and equivalent circuits are of interest. They characterize in a typical manner the devices themselves independent from their implementation in an electric circuit. Whereas the fourpole parameters allow their implementation in the circuit design, those parameters which are not necessarily related to the small signal application are of importance for the device design and are characteristic of the state of the art for a given type of device. In so far as they are related to special devices or at least groups of devices, some of these parameters can also be used generally (*RC* time constants, characteristic frequencies, thermal time constants, contact resistances, etc.). First the transit time and related characteristic frequencies will be considered.

Regarding the devices to be treated in the following sections a selection has to be made because of the large variety of different types of devices. For more insight into the physical principles involved, including those of quantum-effect devices, the reader is referred to Sze (1990) and Shur (1987), the latter covering GaAs devices including transferred electron oscillators and also novel devices (Beneking, 1989). Here their electronic circuit aspects are considered, rather than the physical background of the devices. The important impact of their physics and technology is reflected in the equivalent circuits presented and the formulation of special or general quantities as figures of merit. Taking this into account, conclusions can be drawn regarding the materials to be used for given applications, or the layout and the technology of the devices and integrated circuits to be considered. Unconventional devices are also included, where the principal electric behaviour is reflected in their electronic parameters.

We distinguish here between majority-carrier devices and minority-carrier devices. The first group consists of structures where the overall electric behaviour is determined mainly by the majority carriers in the material, whereas in the second group the minority carriers dominate the behaviour. In the first case diffusion effects are of minor importance. Therefore carrier storage effects or diffusion capacitances can be neglected, leading to fast-reacting devices. Minority-carrier devices can only be similar fast if their active volumes are extremely small, e.g. the base layer in the case of III–V heterojunction bipolar transistors.

5.1 CHARACTERISTIC FREQUENCIES AND TRANSIT TIMES

There are several characteristic frequencies, either related to power relationships or to time constants, relevant for the frequency dependence of device parameters as a figure of merit.

5.1.1 Cutoff frequencies

The term **cutoff frequency** (f_c) is generally used to characterize the frequency where a power transfer is reduced to 50% of its maximum or low frequency value. Because this reduction corresponds to -3 dB on the logarithmic scale, f_c is equal to the **3 dB frequency** ($f_{3\,\mathrm{dB}}$) (e.g. section 4.1.1.). However, the term f_c is also used if other characteristic data are considered.

As a first example a strongly simplified equivalent circuit of a field-effect transistor is discussed (Fig. 5.1(a)). It consists of a gate resistance R in series with the input gate capacitance C_i, and a mutual current source $g_m V$; neglecting an output conductance G_o. Any further components are also suppressed, e.g. the output and load capacitances C_o, C_L.

At a given load (conductance G_L) the magnitudes of the output voltage $|V_L|$ and current $|I_L|$ drop at the cutoff frequency f_c to $2^{-1/2}$ of their initial, low frequency values.

If a simple RC configuration is assumed (onepole configuration, section 4.1.1)

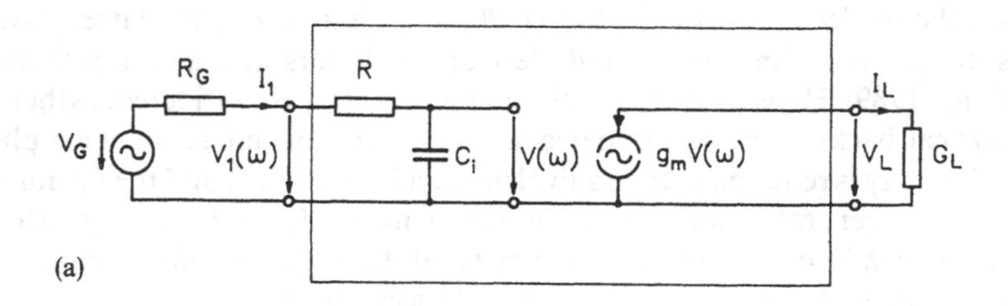

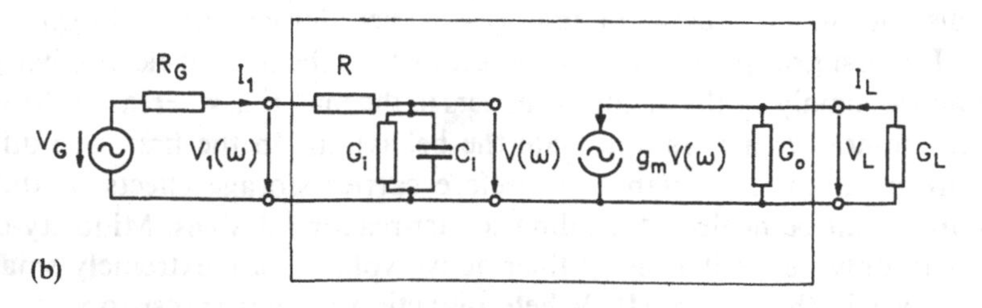

Fig. 5.1 Simplified equivalent circuit of active devices (onepole transfer function). (a) Field effect transistor; (b) bipolar transistor.

the voltage V across the capacitance C_i is

$$V = \frac{V_1}{1 + j\omega RC_i} \tag{5.1}$$

or, in respect of the generator voltage V_G,

$$V = \frac{V_G}{1 + j\omega(R + R_G)C_i} \tag{5.2}$$

Thus the low frequency value of V_L, $f \ll 1/2RC_i$, is

$$V_{L0} = -g_m V R_L = -g_m v_G R_L \tag{5.3}$$

At high frequencies, $f \gg 1/2\pi RC_i$, a frequency-dependent value

$$V_{L,\infty} = j\frac{g_m}{\omega RC_i} V R_L \tag{5.4}$$

and

$$V_{L,\infty} = j\frac{g_m}{\omega(R + R_G)C_i} V_G R_L \tag{5.5}$$

respectively, follows.

The time constant $\tau_i = RC_i$ corresponds to the cutoff frequency

$$f_{ci} = \frac{1}{2\pi\tau_i} \tag{5.6}$$

of the inner device. This is the **45° frequency**, where the imaginary part of the transfer function becomes equal to the real part and the power output becomes one half of its low frequency value. However, if the generator is included and a frequency-independent generator voltage V_G is applied, the corresponding cutoff frequency f_C becomes

$$f_c = \frac{1}{2\pi\tau} \tag{5.7}$$

with

$$\tau = (R + R_G)C_i > \tau_i \tag{5.8}$$

The slope of the magnitude $|V_L| \approx 1/f$ and $P_L(f) \approx 1/f^2 (f \gg f_c)$ is therefore typical for onepole (RC) transfer functions, leading to a -6 dB per octave (frequency ratio 2:1) or −20 dB per decade (frequency ratio 10:1) slope in the log–log representation.

The second example is the circuit of a simplified bipolar transistor, (Fig. 5.1(b)). In discussing the current gain, again the frequency-dependent voltage $V(f)$ across the inner input is responsible for the gain.

Because

$$V = \frac{I_1}{G_i + j\omega C_i} \tag{5.9}$$

the current gain G_I becomes

$$G_I = \frac{I_L}{I_1} = \frac{g_m}{G_i + j\omega C_i} \frac{G_L}{G_L + G_o} \tag{5.10}$$

leading to a power output of

$$P_L = \frac{|I_L|^2}{2G_L} = \frac{|I_1|^2}{2G_i^2} \frac{g_m^2 G_L}{(G_L + G_0)^2} \frac{1}{1 + \omega^2 \tau_i^2} \tag{5.11}$$

Again, one time constant

$$\tau_i = \frac{C_i}{G_i} \tag{5.12}$$

governs the slope of $P_L(f)$, which is, in the case of a conventional bipolar transistor, the diffusion time constant equal to the inner transit time of the device (section 5.2).

Again the corresponding frequency f_{ci} (equation (5.6)) is the cutoff frequency where the power output becomes 50% of its low frequency value at $\omega\tau_i \ll 1$, and also at the same frequency the phase angle of the current gain becomes −45°.

However, this evaluation is only true if the input current I_1 is not frequency dependent. If not $R_G \to \infty$ or at least

$$R_G + R \gg \frac{1}{|G_i + j\omega C_i|} \tag{5.13}$$

then the relevant time constant is modified, because the inner admittance $(G_i + j\omega C_i)$ is shunted by $(R + R_G)$.

This leads to the modified time constant τ instead of τ_i,

$$\tau = \frac{C_i}{G_i + (R + R_G)^{-1}} \tag{5.14}$$

With $R_G \rightarrow 0$ a high f_c-value can be achieved (input, configuration corresponding to Fig. 5.1(b)). Then the time constant is

$$\tau' = \frac{C}{G_i + (1/R)} \tag{5.15}$$

which in practice might be much smaller than τ.

This indicates that a **voltage generator**, $R_G \rightarrow 0$, gives better frequency responses than a **current generator**, $R_G \rightarrow \infty$, if the active device exhibits a configuration corresponding to Fig. 5.1(b), and the overall circuit performance is considered.

A further characteristic frequency can be extracted from the equivalent circuits in Fig. 5.1. With the expression of the short-circuit current gain for the field-effect transistor.

$$G_{Is} = \left.\frac{I_L}{I_1}\right|_{R_L=0} = \frac{-g_m}{j\omega C_i} \tag{5.16}$$

and for the bipolar transistor,

$$G_{Is} = \frac{-g_m}{G_i + j\omega C_i} \tag{5.17}$$

respectively, for both device configurations in the high frequency limit ($\omega \gg G_i/C_i$) follows the general expression

$$|G_{Is}| = \frac{g_m}{\omega C_i} \tag{5.18}$$

Therefore at the **f_1-frequency**

$$f_1 = \frac{g_m}{2\pi C_i} \tag{5.19}$$

the magnitude of the current gain $|G_{Is}|$ becomes unity.

If equal stages are cascaded, then in the high frequency limit also the voltage gain per stage is

$$|G_V| = \frac{g_m}{2\pi f C_i} \tag{5.20}$$

where

$$G_o \ll \left|\frac{1}{R + (1/\omega C_i)}\right| \quad \text{and} \quad G_i \ll \omega C_i \tag{5.21}$$

leading to $|G_V| = 1$ as well as $|G_{Is}| = 1$ at $f = f_1$. As shown in the next section, this characteristic frequency f_1 is equal to the transit frequency f_t, which is related to the transit time in the active device.

5.1.2 Transit times

Transit times t_t are important quantities of high frequency devices, characterizing not only phase lags but also being responsible for the frequency dependence of important device parameters. The transit time is given by

$$t_t = \frac{l}{\bar{v}} \tag{5.22}$$

where l is the length of the region under consideration, and

$$\bar{v} = \frac{1}{l}\int_0^l v(x)\,\mathrm{d}x \tag{5.23}$$

is the mean carrier velocity along l. Correspondingly a transit frequency

$$f_t = \frac{1}{2\pi t_t} \tag{5.24}$$

can be defined, often used to characterize the transit behaviour of an active device. However, f_t does not correspond generally to the cutoff frequency f_c where the transmitted power is one half of its low frequency value.

If a DC current $\bar{I}$ flows continuously through a volume $V = Al$ (Fig. 5.2(a)), the movable electric charge $Q = qnV$ of this volume is exchanged once during the transit time t_t. This follows from

$$\bar{I} = Aqn\,\bar{v} = A\frac{\bar{Q}}{V}\bar{v} = \bar{Q}\frac{\bar{v}}{l} = \frac{\bar{Q}}{t_t} \tag{5.25}$$

where n is the carrier density.

Therefore the transit time can be expressed in a general manner by

$$t_t = \frac{\bar{Q}}{\bar{I}} \tag{5.26}$$

If alternating currents are considered,

$$t_t = \frac{\mathrm{d}\bar{Q}}{\mathrm{d}\bar{I}} \tag{5.27}$$

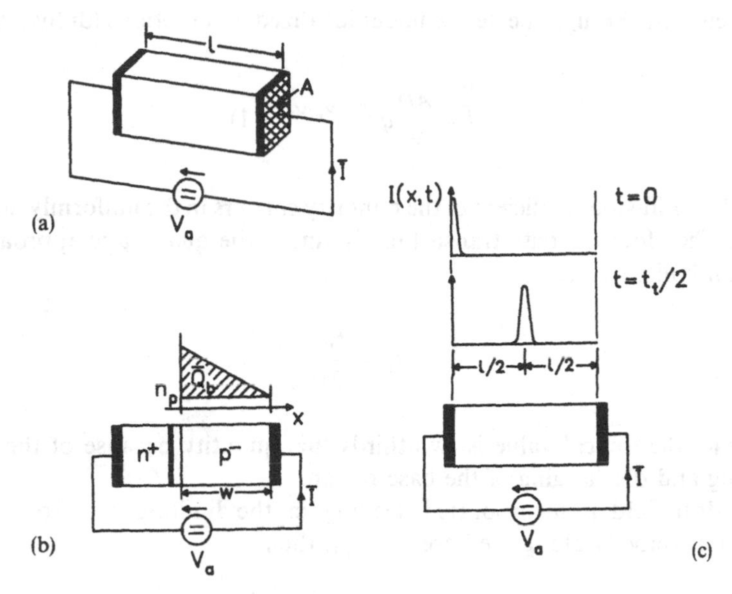

Fig. 5.2 Charge and current. (a) Conventional conductor; (b) bipolar structure; (c) device with induction current.

can be written because of the proportionality of Q and I under the assumption of quasistatic behaviour.

In the case of a conventional bipolar structure (p–n diode, bipolar transistor) the charge to be considered is accumulated in the base region. Assuming a triangular distribution (Fig. 5.2(b)) (no recombination in the base; base width $W \ll L$, the diffusion length of the minority carriers in the base), it follows that

$$\bar{Q}_b = \frac{AW}{2} q n_p (e^{V_a/V_T} - 1) \tag{5.28}$$

where W is the base width, A the area, n_p the minority carrier density of the base and V_a the applied voltage.

The graphical representation in Fig. 5.2(b) could suggest that the charge Q_b of the injected minority carriers would lead to a real (negative) charge in the base. That is not the case as long as extremely high densities are not injected. Majority carriers (holes) compensate this charge by the same amount at any point of the base layer by entering from the contact (on the right hand side). Otherwise high electric fields would be built up by the residual space charge. This charge-compensation occurs very fast, because the relaxation time $t_{relax} = \rho\varepsilon$ of the (highly) doped base layer (ε dielectric constant, ρ specific resistance) is relevant for the majority carriers in contrast to the relatively long diffusion time constant relevant for the minority carriers (section 5.2.1).

The current through the device under idealized assumptions (diffusion current) is

$$\bar{I} = \frac{AD}{W} q n_{\mathrm{p}} (\mathrm{e}^{V_{\mathrm{a}}/V_{\mathrm{T}}} - 1) \tag{5.29}$$

where D is diffusion coefficient of the minority carriers in the uniformly doped base region. Therefore the base transit time is, under the quasistatic approach in the diffusion limited case,

$$t_{\mathrm{tb}} = \frac{W^2}{2D} \tag{5.30}$$

The exact theoretical value is two-thirds this quantity because of the dynamic charging and discharging of the base region.

If a drift field is incorporated, leading in the limiting case to a constant saturation velocity along the base, $v = v_{\mathrm{sat}}$, then

$$t_{\mathrm{tb}} = \frac{W}{v_{\mathrm{sat}}} \tag{5.31}$$

because the carrier distribution becomes nearly constant and any diffusion current can be neglected.

Similarly, along a space charge region, e.g. in the case of a p–n (or MS) junction (Schottky contact) the transit time is

$$t_{\mathrm{ts}} = \frac{d}{v_{\mathrm{sat}}} \tag{5.32}$$

because of the high electric field in the depleted zone (length d), leading to velocity saturation.

The latter generally corresponds to the flux of majority carriers, if saturation occurs, otherwise the transit time is given by the first general expression (equation (5.22)).

As mentioned in the foregoing section the transit frequency f_{t} is equal to the frequency f_1 where the magnitude of the short-circuit current gain of an active device becomes unity. This generally follows from

$$\frac{1}{2\pi f_{\mathrm{t}}} = t_{\mathrm{t}} = \frac{\mathrm{d}\bar{Q}}{\mathrm{d}\bar{I}} = \frac{\mathrm{d}\bar{Q}}{\mathrm{d}V}\frac{\mathrm{d}V}{\mathrm{d}\bar{I}} = C_{\mathrm{i}}\frac{1}{g_{\mathrm{m}}} = \frac{1}{2\pi f_1} \tag{5.33}$$

where V is the input voltage applied to control the output current, $\bar{I}$, C_{i} and g_{m} having been defined in section 5.1.1. If the time behaviour of the applied signals becomes comparable to the transit time, or the frequency is not low with respect to f_t, then the quasistatic approximation fails, and the induction current following from the carrier behaviour in the volume under consideration has to be evaluated.

The induction current I_i is the mean current in the outer circuit, given by

$$I_i = \frac{1}{l}\int_0^l I(x, t)\,dx \tag{5.34}$$

if $I(x, t)$ is the convection current in the drift region with length l. If at $x = 0$ a current pulse is injected according to the schematic drawing in Fig. 5.2(c), represented by

$$i = I_0\, e^{j\omega t} \tag{5.35}$$

then

$$I(x, t) = I_0 \exp\left(j\omega\left(t - \frac{x}{v_{sat}}\right)\right) \tag{5.36}$$

if this signal travels as a charge packet along x with the saturation velocity

$$v_{sat} = \frac{l}{t_t} \tag{5.37}$$

After integration it follows that

$$I_i = \frac{I_0\, e^{j\omega t}}{j\omega t_t}(1 - e^{-j\omega t_t}) \tag{5.38}$$

or

$$I_i = i\, e^{-j(\omega t_t/2)} \left\{ \frac{\sin\left(\frac{\omega t_t}{2}\right)}{\frac{\omega t_t}{2}} \right\} \tag{5.39}$$

At low frequencies, $\omega t_t \ll 1$,

$$I_i = i \tag{5.40}$$

however, at higher frequencies a delay occurs equal to half the transit time, and the amplitude becomes reduced.

At $\omega t_t/2 = 1.39$ the amplitude has dropped to $2^{-1/2}$ of its low frequency quasistatic value, leading to the 3 dB cutoff frequency

$$f_c = \frac{0.44}{t_t} \tag{5.41}$$

Because the transit frequency is

$$f_t = \frac{1}{2\pi t_t} = \frac{0.16}{t_t} \tag{5.42}$$

it follows that

$$f_c \approx 2.75 f_t \tag{5.43}$$

This value is roughly a factor of three higher than the corresponding transit frequency.

If a pulse transmission is considered, correct transmission of at least the third harmonics is necessary. Therefore f_t has to be about equal to the maximum repetition rate R_{max} which is a design consideration, e.g. for photodetectors for optoelectronic communication at high bit rates.

5.2 DIODES

Any conventional active device is composed of diodes. However, the diodes themselves exhibit special features. In the following a short review of their principal behaviour is given, and related characteristic electrical data are derived.

Quantum well structures can also exhibit diode-like characteristics, e.g. resonant tunnelling diodes or gradual gap structures. Those elements are not considered here in detail. From an application point of view these devices and also modulation-doped barriers such as Camel diodes or triangular barriers can be treated in a similar way as conventional devices (e.g. Beneking, 1989). Devices with partial negative slope in their DC characteristics are treated in section 5.5 after a description of the tunnel diode in section 5.2.2.

5.2.1 P–n diodes and M–S diodes

The static characteristic of a diode is given by the general equation

$$I_P = I_0(e^{V_B/\nu V_T} - 1) \tag{5.44}$$

or

$$I_P = I_0\left(\exp\left(\frac{V_P - R_b I_P}{\nu V_T}\right) - 1\right) \tag{5.45}$$

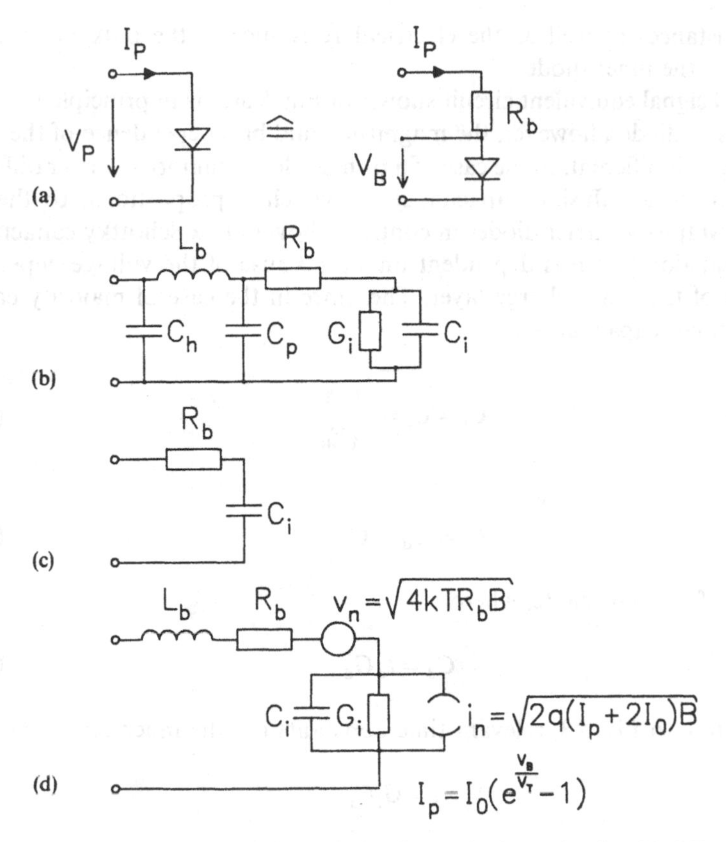

Fig. 5.3 Diode equivalent circuits. (a) Static equivalent circuit; (b) RF small signal equivalent circuit; (c) simplified small signal equivalent circuit; (d) incorporation of noise sources (B is the bandwidth).

with the inner DC voltage across the junction (Fig. 5.3(a)).

$$V_B = V_P - R_b I_P \tag{5.46}$$

This nonlinear function is relevant for the rectification properties of a conventional diode.

The coefficient $\nu = 1$–2 is the ideality factor, being 1 in case of a perfect p–n junction and 2 in case of high injection and recombination in the space charge region, respectively. Majority carrier diodes (Schottky diodes, Camel diodes, triangular barriers) exhibit $\nu \approx 1.5$, and the reverse current I_0 is voltage dependent and about three orders of magnitude larger than in the case of a p–n diode. Therefore ν and $J_0 = I_0/A$ at a given reverse voltage are figures of merit of a diode, besides the breakdown voltage in reverse direction, and the base resistance R_b (A is the diode area). The latter represents parasitic series resistances and contact resistances as well as the electrical resistance of the bulk material on both sides of the inner diode.

The small signal equivalent circuit shown in Fig. 5.3(b) is in principle the same for all types of diodes, however, the magnitude and bias dependence of the inner capacitance C_i is different. In the case of a p–n diode the minority carrier diffusion current leads to a diffusion capacitance C_d which is proportional to the DC current I_P. Majority carrier diodes in contrast show only a Schottky capacitance C_s at the junction which is dependent on V_B because of the voltage dependent thickness d of the space charge layer. Therefore in the case of majority carrier diodes the inner capacitance is

$$C_i = C_s = \frac{\varepsilon_s A}{d(V_B)} \tag{5.47}$$

instead of

$$C_i = C_d + C_s \tag{5.48}$$

in the case of the p–n diode, where

$$C_d = t_d G_d \tag{5.49}$$

Corresponding to Fig. 5.3, relevant time constants are the inner time constant

$$\tau_i = G_i C_i \tag{5.50}$$

being often negligible in case of majority carrier diodes, and the time constant

$$\tau_s = R_b C_s \tag{5.51}$$

which is of importance if the diode is reverse biased, $V_B < 0$. This time constant does not include C_d because in the case of a p–n diode the diffusion capacitance C_d becomes zero in the reverse direction.

A special class of diodes are Mott diodes which are Schottky diodes with a totally depleted base region, at least for reverse applied voltages. This leads to a constant, voltage-independent value of C_s, and between the Schottky contact and the metallic base contact a high ohmic zone with a constant electric field is built up. This allows their application as very fast photodetectors (e.g. Beneking, 1989).

The different physical origin of C_d and C_s is relevant for the different electronic behaviour of both capacitances, not only for their small signal behaviour but also for large signal switching. The Schottky capacitance C_s is a capacitance equivalent to the common capacitance formed by two plates, separated by the dielectric. In semiconductor material the dielectric is the space charge layer, which is depleted more or less of movable carriers (electrons and holes).

In the case of a p–n junction the plates correspond to the neutral p and n regions on both sides of it. In the case of a Schottky capacitance the plates are on the one side the metal and on the other side the residual neutral, doped semiconductor corresponding to that of the p–n junction – providing that reach through does not occur, as in case of a Mott diode.

Therefore the charging and discharging occurs at both interfaces by sweeping in and out movable majority carriers from outside. At the semiconductor sites this becomes possible via the varying thickness of the depleted region. The effect is therefore related to the majority carriers on both sides of the dielectric, and up to the relaxation frequency $f_{relax} = 1/\varepsilon\rho$ (ε dielectric constant, ρ specific resistance of the semiconductor material) no frequency dependence exists.

Contrary to this the diffusion capacitance C_d is principally frequency dependent (section 1.3.1). There both movable charges are stored together in the same volume, because point by point charge neutrality has to be established. The electrons flow in from the one side, the holes in the same amount from the other, both arranging in the same spatial distribution, dependent on the applied voltage (section 5.1.2).

This underlines that the equivalent circuit with constant, frequency independent components as well as the given static characteristics, are assumptions, representing the diode sufficiently but not exactly. The resistance R_b, for example, is in reality a distributed structure like a transmission line, leading to a frequency dependence which is commonly neglected.

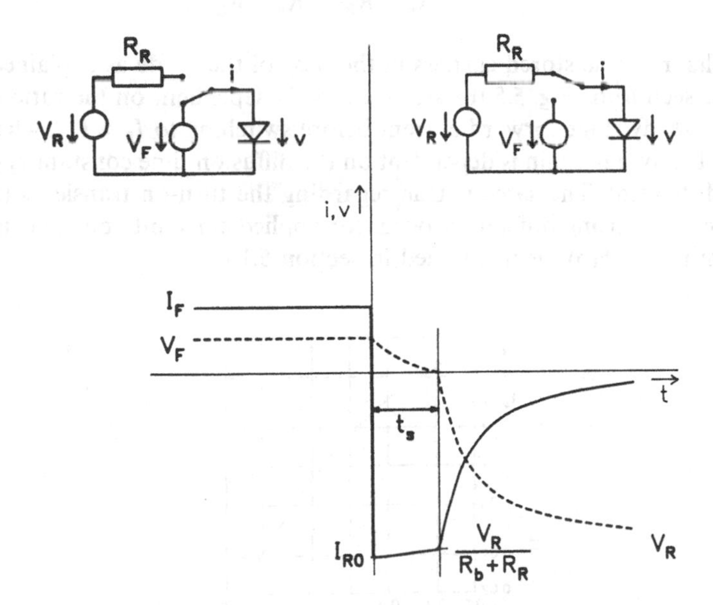

Fig. 5.4 Large signal transients of a p–n diode. Indicated are the current (——) and the voltage (- - -) across the diode in the case of turn-off from a conducting state. The inset shows the circuit configuration.

If a diode is used as a large signal switching element, again the given time constants are of interest. In the case of a majority carrier diode no storage time t_s exists, only τ_s is relevant for the switching behaviour. No charge is built up in the non-depleted base region while the current flow is in the 'on' condition, which should first be extracted. However, a p–n diode shows a storage time t_s, because the charge $\overline{Q_b}$ stored in the base (Fig. 5.2(b)), has to be removed before the 'off' condition with low reverse current I_0 can be reached (Beneking, 1965; Illi, 1968; Kingston, 1954; Le Can *et al.*, 1962). During that time interval the electrons move out in the direction from where they came, correspondingly the holes move back to the other side. One part of both carriers recombines directly in the volume itself, which part is dependent on the given configuration and the applied reverse voltage (section 5.1.2).

Similar conclusions can be drawn in the case of bipolar transistors. These allow the application of a Schottky diode as the collector part of the device achieving negligible storage at the collector junction and therefore $t_s = 0$ (Schottky collector transistor, Fig. 6.2).

Figure 5.4 shows the time-dependent voltage and current if a p–n diode is switched off from an on state (also Fig. Ex6.1). In the first moment a reverse current $I_{RO} \gg I_0$ occurs which for $|V_R| \gg V_F$ is determined alone by the applied reverse voltage V_R and the series resistance $R_b + R_R$,

$$I_{RO} = \frac{|V_R| + V_F}{R_b + R_R} \approx \frac{|V_R|}{R_b + R_R} \tag{5.52}$$

It results from the stored carriers in the base of the diode as explained above. As can be seen from Fig. 5.5 the storage time is dependent on the ratio I_R/I_F, I_F being the steady stage forward current before switching to $I_R \approx I_{RO}$, whereas the principal behaviour again is dependent on the diffusion time constant t_d (without built-in drift field). The same is true regarding the turn-on transients (Fig. 5.6). As can be seen, strong differences occur for applied constant voltage (Fig. 5.6(a)) or current (Fig. 5.6(b)) as mentioned in section 5.1.1.

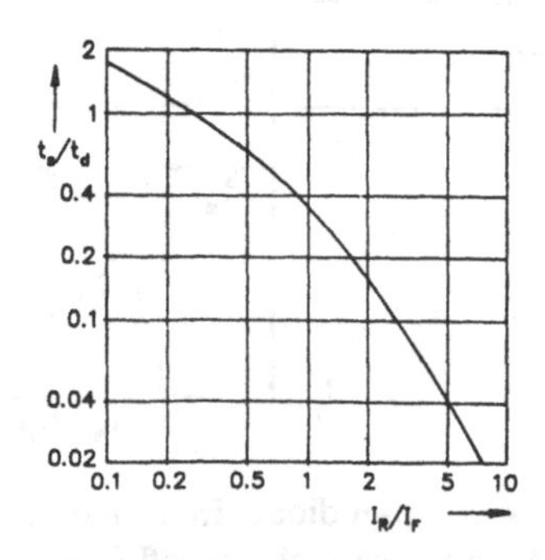

Fig. 5.5 Storage time t_s of a p–n diode normalized to $t_d = W^2/2D$). (After Kingston, 1954.)

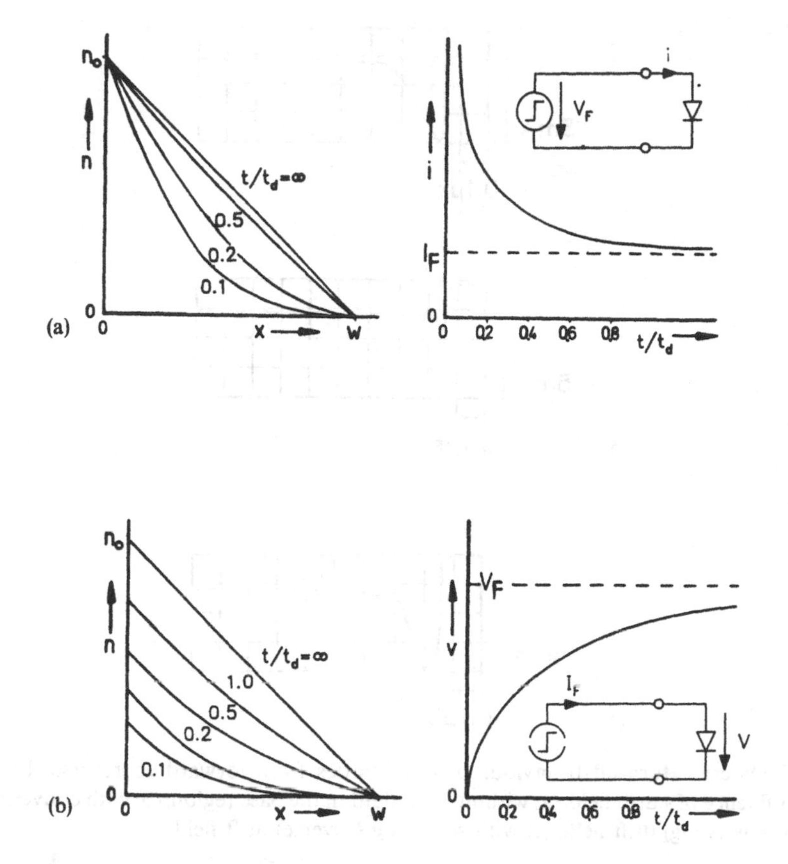

Fig. 5.6 Turn-on behaviour of a p–n diode. The minority carrier density $n(x)$ ($W \ll L$ diffusion length) and the electrical transient ($t_d = W^2/2D$) are shown.

The decay behaviour after t_s is dependent on the residual charge distribution in the base. If a retarding drift field is built in, a sharp turn-off occurs because of the accumulation of the residual charge near the junction. This allows for effective frequency multiplication (step recovery diode). On the other hand a conventional drift field in the forward direction slows down the decay. This is demonstrated in Fig. 5.7, where the large signal switching behaviour is shown for otherwise identical diodes, exhibiting a retarding drift field (c) and a normal drift field (b). In the first case the extremely sharp transient can be seen whereas the positive drift field as used in the drift transistor leads to a reduced tail. Regarding the decay behaviour of p–i–n diodes with low n-type doping in the middle zone (p–s_n–n diodes) or low p-type doping (p–s_p–n diodes) differences also exist. This is shown in Fig. 5.8.

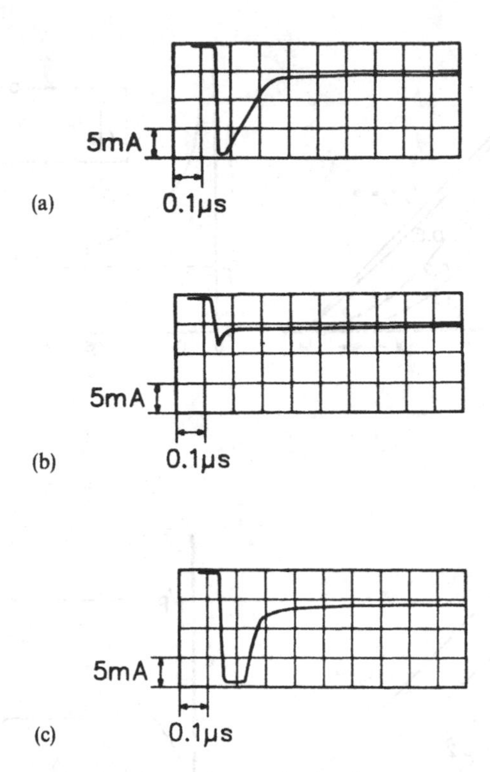

Fig. 5.7 Different turn-off behaviour of p–n diodes from forward to reverse direction under influence of a drift field: (a) without drift field in the base region; (b) with conventional (diffusion assisting) drift field; (c) with retarding (reverse) drift field.

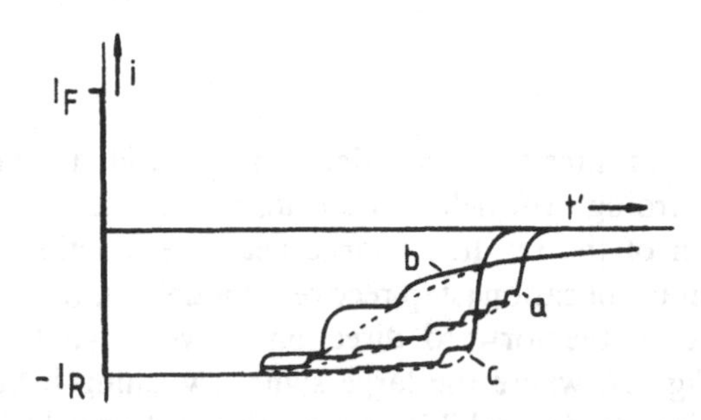

Fig. 5.8 Large signal turn-off behaviour of p–n diodes (analog model simulation). (a) Middle zone intrinsic; (b) middle zone lightly n-type doped; (c) middle zone lightly p-type doped. (After Illi, 1968.)

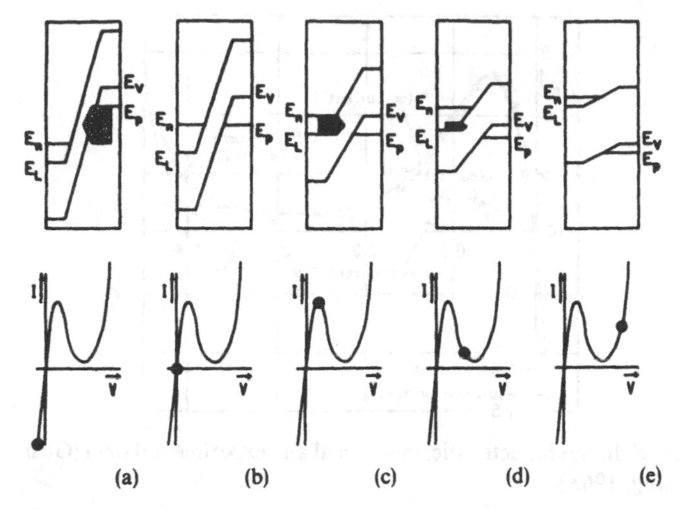

Fig. 5.9 Behaviour of an n^+–p^+ tunnel diode, band diagram and electrical characteristics. (a) Zener breakdown in reverse direction; (b) zero bias condition; (c) peak point: electron tunnelling into the valence band; (d) negative conduction region (near valley point); (e) diffusion regime.

5.2.2 Tunnel and backward diodes

A further class of diodes are the tunnel diode and the backward diode (e.g. Scanlan, 1966). Those diodes exhibit static characteristics which are dependent on the tunnelling of electrons between the conduction band and the valence band, if the diode is degenerately doped (Fig. 5.9). This effect results in a differential negative conductance (Fig. 5.9(d)) or, depending on the technology, in a sharp knee for high rectification efficiency (Fig. 5.9(c)). The general properties of those negative resistance devices are treated in section 5.5.

The static characteristic of the tunnel diode is given according to Vogelsang (1963) by

$$I = I_0(e^{V/V_T} - 1) + I_p \frac{V}{V_p} \exp\left(1 - \frac{V}{V_p}\right) + G_e V \tag{5.53}$$

Characteristics of other negative impedance devices can be similarly expressed.

The corresponding current components of a tunnel diode are indicated in Fig. 5.10. Differentiating this expression gives the maximum negative conductance $-|G_N|_{max}$ achievable, and its bias condition. It follows that

$$|G_N|_{max} = \frac{I_p}{eV_p} \tag{5.54}$$

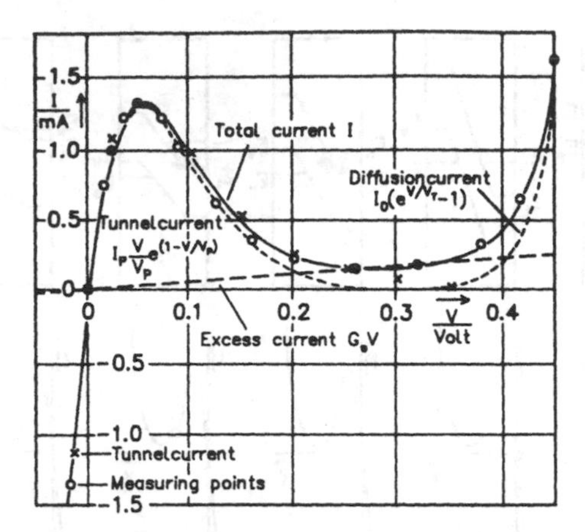

Fig. 5.10 Tunnel diode characteristic, theoretical and experimental data (Ge $n^{+}-p^{+}$ diode). (After Vogelsang, 1963.)

and

$$I|_{|G_N|_{max}} = \frac{2}{e} I_p \tag{5.55}$$

The equivalent circuit is the same as in the case of a conventional semiconductor diode, because these diodes are p–n diodes degenerately doped to narrow the space charge layer. Because their application is restricted to the region around the peak point at low forward voltage the diffusion capacitance C_d can generally be neglected, and only $\tau_s = R_b C_s$ is relevant (Fig. 5.3). Therefore these diodes are applicable up to relatively high frequencies in the GHz range.

Resonant tunnel structures exhibit similar DC characteristics with partial negative slope. They are symmetrical majority carrier devices without diffusion capacitances where the negative conductance becomes pronounced at low temperatures (liquid nitrogen). The transit times are again shorter as in the case of the heavily doped tunnel diode because of the extremely thin layer sequence (in total $<0.2\,\mu m$). This enables us to apply these configurations up to THz frequencies (Sollner *et al.*, 1983).

However, because of parasitic components, stability problems exist which impede the biasing of a tunnel diode or other negative conductance devices. First of all the DC resistance $R = R_L + R_S$ has to be smaller than the negative resistance at the inner bias point D (Fig. 5.11(a)),

$$\frac{1}{R} > |G_N| \tag{5.56}$$

to be able to fix the bias point. Otherwise the DC load line $R'_L + R_S$ instead of

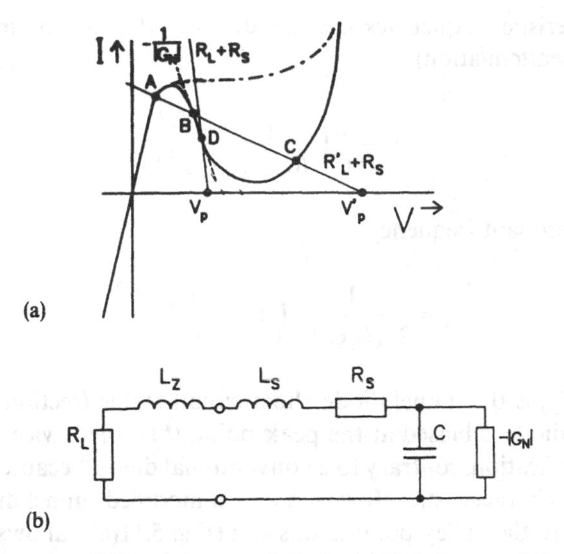

Fig. 5.11 Biasing and stability of the tunnel diode. (a) DC characteristics (·–·–·, backward diode); (b) RF equivalent circuit.

$R_L + R_S$ crosses the diode characteristic at the points A, B and C, where only A and C are stable.

On the other hand, because of resonance transformation of the impedances via the L–C configuration (Fig. 5.11(b)) RF stability along the negative branch only exists for

$$\frac{|G_N|}{C} < \frac{R_1}{L_Z + L_S} \tag{5.57}$$

In this equation R_1 is the acting RF resistance

$$R_1 = R_S + R_L(\mathrm{RF}) \tag{5.58}$$

Therefore a window exists in respect of the ohmic load and the DC bias resistance R_L, respectively, for stable operation,

$$\frac{1}{|G_N|} > R > \frac{(L_Z + L_S)|G_N|}{C} \tag{5.59}$$

assuming $R_1 = R = R_L + R_S$. Only under this condition can the tunnel diode or a resonant tunnelling barrier be used as a negative conductance amplifier or oscillator.

Two characteristic frequencies can be derived, the maximum frequency of amplification (deattenuation)

$$f_{c0} = \frac{|G_N|}{2\pi C}\left(\frac{1}{R_S|G_N|} - 1\right)^{1/2} \tag{5.60}$$

and the inner resonant frequency

$$f_{os} = \frac{1}{2\pi(L_S C)^{1/2}}\left(1 - \frac{L_S G_N^2}{C}\right)^{1/2} \tag{5.61}$$

Only for $f_{os} > f_{c0}$ is the tunnel diode short-circuit-stable (section 5.5).

If the tunnel diode is biased at the peak point, then the device can be used as mixer with amplification, contrary to a conventional diode, because of the positive curvature of the characteristic. However, also a modified tunnel diode, the backward diode where the valley point is missing (Fig. 5.11(a)), allows effective conversion and rectification. In this case no negative conductance occurs. However, the curvature at the residual peak point is equally high.

Further, two-terminal devices exist especially for GHz applications (Impatt diodes, the Gunn element, etc.). These well documented devices are not covered here (e.g. Bosch and Engelmann, 1975; Shur, 1987; Sze, 1990).

5.2.3 Noise parameters

Referring to the physical noise sources listed in Appendix F the equivalent noise circuit of a diode can be simply derived. Besides the contribution of ohmic resistances and generation–recombination effects in the space charge layer (or at the M–S interface in case of Schottky diodes) shot noise does occur related to the carrier drift through the field region of the junction.

With the static characteristics of an ideal diode

$$I = I_0(e^{V/V_T} - 1) \tag{5.62}$$

two such currents have to be considered, $I_0 e^{V/V_T}$ in the forward direction and I_0 in the reverse direction. Therefore shot noise is involved corresponding to

$$d(i_{ns})^2 = 2qI_0 e^{V/V_T} df + 2qI_0 df \tag{5.63}$$

leading to

$$d(i_{ns})^2 = 2q(I + 2I_0) df \tag{5.64}$$

Only for $V \neq 0$ is the diode under nonthermal conditions and shot noise occurs. At $V = 0$ the inner diode acts as a passive device and shows thermal noise. This follows from the equation above with $I = 0$ and the given dynamic conductance

$$G = \frac{\mathrm{d}I}{\mathrm{d}V} = \frac{I_o}{V_T} e^{V/V_T} = G_o e^{V/V_T} \tag{5.65}$$

At $V = 0$ it is

$$G_o = \frac{qI_0}{kT} \tag{5.66}$$

and

$$\mathrm{d}(i_n)^2 = 4qI_0\mathrm{d}f = 4kTG_o\mathrm{d}f \tag{5.67}$$

Figure 5.3(d) shows the simplified noise equivalent circuit after Fig. 5.3(b).

In the breakdown region (reverse direction) additional noise effects occur. Two predominant effects have to be mentioned. If extremely high doping exists, leading to a short depletion zone, very high electric fields occur for small voltages (a few volts). Then tunnelling of electrons and/or holes through the junction is the dominant current origin, and Zener breakdown occurs. Because single carriers cross the junction independently, the conventional shot noise formula can be applied (section F.2). For lower doping the breakdown voltage rises, and the current in the reverse direction results from the statistical turn on and off of microplasma in the high field zone of the junction. It follows for the spectral density of the noise voltage source (Haitz, 1967) that

$$\mathrm{d}(v_n)^2 = a^2 \frac{V_{Br}}{\bar{I}_R} \frac{1}{\left\{1 - \left(\frac{f}{f_A}\right)^2\right\}^2} \mathrm{d}f \tag{5.68}$$

where $\bar{I}_R$ is the mean reverse current and V_{Br} is the breakdown voltage. The factor a^2 is a material dependent quantity. For silicon $a^2 \approx 3.3 \times 10^{-20}\,\mathrm{A\,Hz^{-1}}$. The avalanche frequency f_A is relevant for the switching speed of the microplasma, $f_A \approx 1 \cdots 10\,\mathrm{GHz}$.

5.2.4 Zener diodes and noise diodes

Zener diodes are p–n diodes where a sharp knee of the DC characteristic in reverse direction allows for stabilization of DC voltages if their bias point is adjusted to the breakdown branch of the characteristic (Fig. 5.12).

In Fig. 5.12(a) the common stabilization circuit is indicated. Figure 5.12(c) shows the equivalent voltage source, whereas the graphical determination of the residual variation of the output voltage is indicated in Fig. 5.12(d), (e).

For effective stabilization a small dynamic resistance R_{Br} of the breakthrough

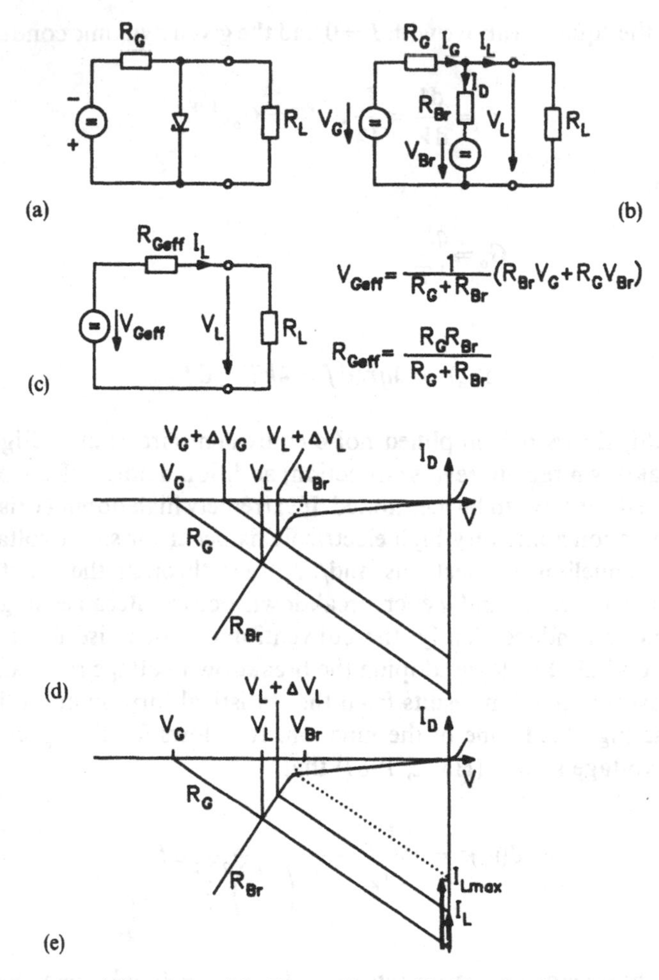

Fig. 5.12 Voltage stabilization with Zener diode. (a) Circuit; (b) equivalent circuit; (c) effective voltage source; (d) I–V diagram ($I_L = 0$); (e) I–V diagram ($I_L \neq 0$).

characteristic at the bias point is decisive (Fig. 5.12(d). Correspondingly the effective generator resistance is reduced, $R_{Geff} \approx R_{Br}$, (Fig. 5.12(e)),

The internal physical effect of breakdown can be either field emission (Zener effect) or impact ionization (avalanche effect). Avalanche diodes show in general a sharper knee than real 'Zener' diodes. Each type exhibits a different temperature behaviour, as shown in Fig. 5.13. This leads to temperature independent stabilization for $|V| \approx 6\,V$. (For the stabilization of small voltages, $|V| > V_D \approx 1\,V$, forward-biased diodes can be used.)

As mentioned at the end of the foregoing section, avalanche diodes exhibit in the (low ohmic) avalanche breakdown regime microplasma noise. The correspondent noise power can exceed the thermal noise of an equivalent resistor

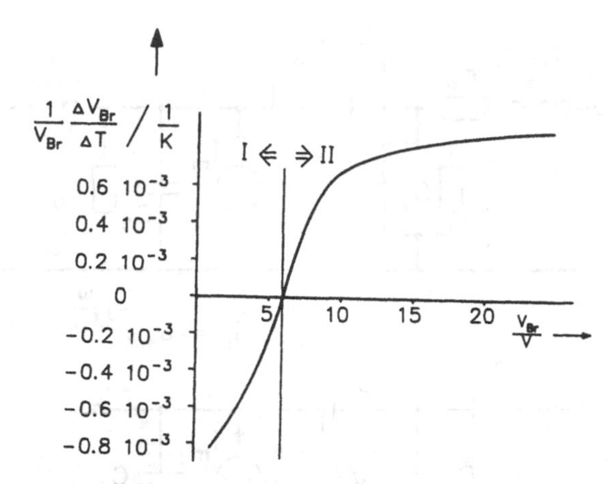

Fig. 5.13 Temperature coefficient of Zener diodes (T_j junction temperature; region I Zener breakdown, region II avalanching.)

($R = R_{Br}$) by a factor of 10 000 (40 dB). Because additionally the generated noise is (about) white noise up to GHz frequencies those diodes are applicable as stable RF noise sources for noise measurements (e.g. Tsironis and Beneking, 1976).

5.3 BIPOLAR TRANSISTORS

In Chapter 2, and sections 5.1.1 and 5.1.2, general parameters have been introduced to characterize active devices. This section relates them specifically to the bipolar transistor, referring to a more exact equivalent circuit as introduced in section 5.1.1 (Fig. 5.1(b)).

5.3.1 Electronic parameters

In Fig. 5.14 the extended circuit is shown. It consists of a modified Π equivalent circuit and includes important parasitic components, first derived by Giacoletto (1954).

The accuracy obtainable in computing the parameters from measured data has been demonstrated in section 1.3.2 (Figs. 1.8 and 1.15).

A simplified version can be used at higher frequencies (Fig. 5.14(b)). However, regarding the neglected phase angle of the transconductance, $\varphi \approx \omega/3\omega_t$, $\omega_t = 2\pi f_t$ (f_t transit frequency), it can be used only as a rough estimate. As mentioned in section 5.2 all equivalent circuits with constant, frequency-independent components are assumptions. However, they represent well the device under consideration as long as the equivalent circuit is sufficiently complicated to include the major physical effects.

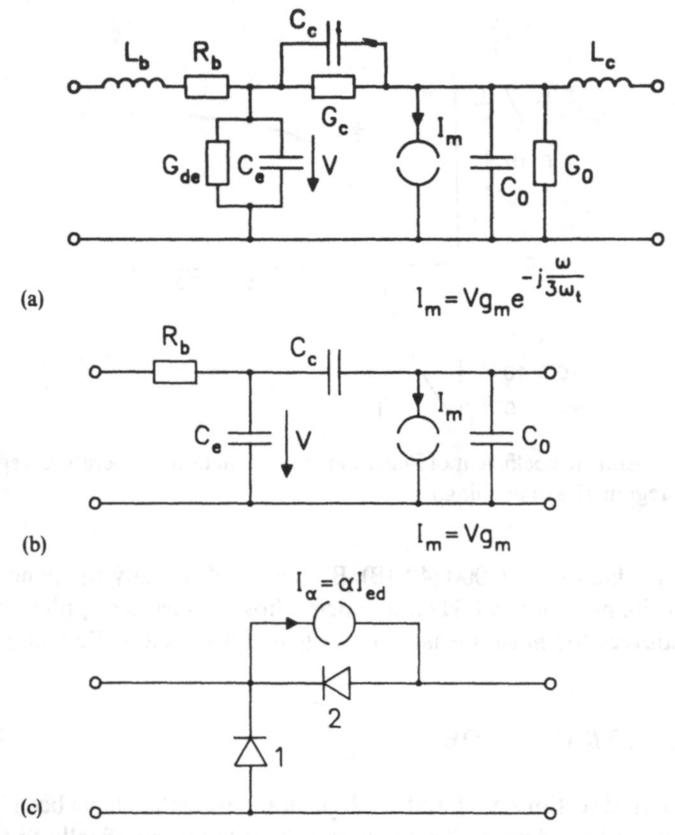

Fig. 5.14 Equivalent circuit of a bipolar transistor (grounded emitter) (a) Complex equivalent circuit with parasitic components ($C_e = C_{de} + C_{se}$, $C_c = C_{dc} + C_{sc}$); (b) simplified high frequency equivalent circuit; (c) T equivalent circuit, simplified version (diode 1 forward biased, diode 2 reverse biased, I_{ed} diffusion current of diode 1, α common base current amplification factor).

Instead of the Π equivalent circuit the *z*-parameter-related *T* equivalent circuit can also be used. Figure 5.15 gives an example (Bayraktaroglu and Camilleri, 1988). The quasistatic version (Fig. 5.14(c)) is well adapted for large signal evaluations including CAD.

The simplified equivalent circuit shown in Fig. 5.14(b) allows us to determine f_t and f_{max} for the bipolar transistor. The *y*-parameters are, taking into account the high frequency limit $\omega R_b C_e \gg 1$ and assuming $C_e \gg C_c, g_m > \omega C_c$

$$y_{11} = \frac{j\omega C_e}{1 + j\omega R_b(C_e + C_c)} \approx \frac{1}{R_b} \tag{5.69}$$

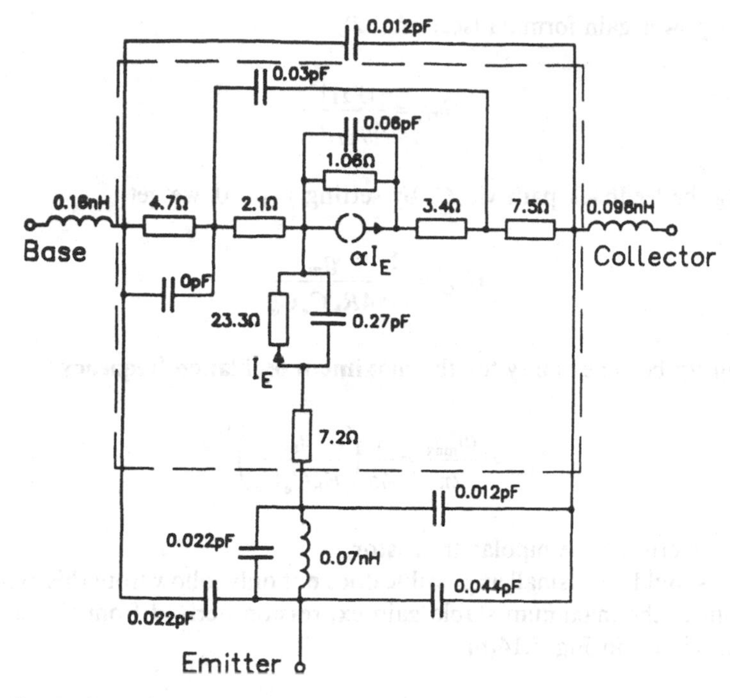

Fig. 5.15 Equivalent circuit and element values for 60 µm emitter periphery p–n–p GaAlAs/GaAs MOVPE grown heterojunction bipolar transistor ($f_t = 12\,\text{GHz}$, $f_{max} = 20\,\text{GHz}$; $w_b = 0.1\,\mu\text{m}$, $\beta_0 = 60$; $\alpha = \alpha_0 e^{-j\omega\tau}/[1 + j(f/f_\alpha)]$($\alpha_0 = 0.96$; $\tau = 4\,\text{ps}$; $f_\alpha = 35\,\text{GHz}$). (After Bayraktaroglu and Camilleri, 1988.)

$$y_{12} = -\frac{j\omega C_c}{1 + j\omega R_b(C_e + C_c)} \approx \frac{-C_c}{R_b C_e} \tag{5.70}$$

$$y_{21} = \frac{g_m - j\omega C_c}{1 + j\omega R_b(C_e + C_c)} \approx \frac{-jg_m}{\omega R_b C_e} \tag{5.71}$$

$$y_{22} = j\omega C_o + j\omega C_c \frac{\dfrac{1}{R_b} + j\omega C_e}{\dfrac{1}{R_b} + j\omega(C_e + C_c)} + g_m \frac{j\omega C_c}{\dfrac{1}{R_b} + j\omega(C_e - C_c)}$$

$$\approx j\omega(C_o + C_c) + g_m \frac{C_c}{C_e} \tag{5.72}$$

Using the power gain formula (section 2.2),

$$G_{\mathrm{TO}} = \frac{|y_{21}|^2}{4g_{11}g_{22}} \tag{5.73}$$

neglecting the feedback path via C_c by setting $y_{12} = 0$, we get

$$G_{\mathrm{TO}} = \frac{1}{\omega^2} \frac{g_m}{4R_b C_e C_c} \tag{5.74}$$

This quantity becomes unity for the **maximum oscillation frequency**

$$f_{\max} = \frac{\omega_{\max}}{2\pi} = \frac{1}{4\pi}\left(\frac{g_m}{R_b C_e C_c}\right)^{1/2} \tag{5.75}$$

a figure of merit for the bipolar transistor.

That C_c should be as small as possible does not only follow from this expression but also from the maximum stable gain expression derived from the simplified equivalent circuit in Fig. 5.14(b).

With

$$\mathrm{MSG} = \left|\frac{y_{21}}{y_{21}}\right| = \frac{g_m}{\omega C_c} \tag{5.76}$$

the gain at the stability limit as inversely proportional to C_c. Using the y-parameter derived above the short-circuit current gain $h_{21} = y_{21}/y_{12}$ of the transistor is

$$h_{21} = -\frac{jg_m}{\omega C_e} \tag{5.77}$$

The magnitude $|h_{21}|$ of this quantity becomes units at

$$\omega_1 = \frac{g_m}{C_e} \tag{5.78}$$

which defines the **f_1-frequency** (section 5.11)

$$f_1 = \frac{\omega_1}{2\pi} = \frac{g_m}{2\pi C_e} \tag{5.79}$$

This frequency can easily be determined by measuring $|h_{21}|$ at a frequency f_m

in the region where $|h_{21}|$ is inversely proportional to the frequency. Then

$$f_1 = |h_{21}||_{f_m} f_m \tag{5.80}$$

Introducing f_1 into the formula for f_{max}, we get

$$f_{max} = \frac{1}{4\pi}\left(\frac{2\pi f_1}{R_b C_c}\right)^{1/2} \approx \frac{1}{5}\left(\frac{f_1}{R_b C_c}\right)^{1/2} \tag{5.81}$$

Therefore besides f_1 the RC product $R_b C_c$ is also a figure of merit; the smaller it is the higher the frequency limit of the device. The quantity f_1 corresponds to the transit frequency $f_t = 1/2\pi t_t$, where t_t is the signal transit time in the inner device (section 5.1.2).

The transit time along the base width W is (equation (5.30))

$$t_{tb} = \frac{W^2}{2D}$$

in the case of a diffusion-governed transport of the minority carriers through the base and is identical with the diffusion time constant t_d (diffusion coefficient D). If a drift field is incorporated leading to a mean (saturation) velocity $\bar{v}$ in the base, then in the limiting case

$$t_t = \frac{W}{\bar{v}} \tag{5.82}$$

is achieved. Setting

$$\frac{W}{\bar{v}} = \frac{W^2}{2D} \tag{5.83}$$

with $\bar{v} = v_{max}$, we can calculate the minimum base width W_{min} for which the implementation of a drift field enhances the carrier velocity in respect of the diffusion case. It follows that for

$$W < W_{min} = \frac{2D}{v_{max}} \tag{5.84}$$

saturation occurs.

The transit time can be very short, of the order of 1 ps. For example for $D = 100\,cm^2\,s^{-1}$ and $v_{max} = 10^7\,cm\,s^{-1}$, $W_{min} = 0.2\,\mu m$, and t_t becomes 2 ps. Then the transit times along the space charge layers (emitter–base, base–collector), thicknesses d_{EB}, d_{BC}, can no longer be neglected, and t_{tmin} becomes

$$t_{\mathrm{tmin}} = \frac{W + d_{\mathrm{EB}} + d_{\mathrm{BC}}}{v_{\mathrm{max}}} \tag{5.85}$$

Both frequencies $f_t = f_1$ and f_{max} are the most commonly used figures of merit. In practice they are determined by linear extrapolation of the measured or computed values of $|h_{21}|$ and G_{T} on a log–log scale which of course involves the uncertainty of the real slope at higher frequencies for the measured values (e.g. Fig. 2.4).

As pointed out in section 5.1.1, f_1 not only characterizes the limiting frequency for the current transfer. It also indicates generally the frequency where, by cascading the same active device, the voltage gain drops to unity.

Using the y_{kl} parameters derived above the voltage gain

$$G_{\mathrm{v}} = \frac{-y_{21}}{y_{22} + Y_{\mathrm{L}}} \tag{5.86}$$

becomes

$$G_{\mathrm{v}} = \frac{-g_{\mathrm{m}}}{\mathrm{j}\omega R_{\mathrm{b}} C_{\mathrm{e}}} \frac{1}{\mathrm{j}\omega(C_{\mathrm{o}} + C_{\mathrm{c}}) + g_{\mathrm{m}}\dfrac{C_{\mathrm{c}}}{C_{\mathrm{e}}} + Y_{\mathrm{L}}} \tag{5.87}$$

If

$$|Y_{\mathrm{L}}| \approx \frac{1}{R_{\mathrm{b}}} \gg \omega(C_{\mathrm{o}} + C_{\mathrm{c}}), g_{\mathrm{m}}\frac{C_{\mathrm{c}}}{C_{\mathrm{e}}} \tag{5.88}$$

the magnitude of G_{v} becomes

$$|G_{\mathrm{v}}| = \frac{g_{\mathrm{m}}}{\omega C_{\mathrm{e}}} \tag{5.89}$$

which is equal to the current gain $|h_{21}|$ in the upper frequency limit.

An equivalent conclusion can be drawn for the case of a cascaded FET amplifier (section 5.4.1).

5.3.2 Noise parameters

Using the equivalent circuit of Fig. 5.14(c) the noise sources relevant for a bipolar transistor can easily be implemented. Referring to the noise sources listed in Appendix G it follows that besides noise sources belonging to parasitic components (e.g. emitter resistance, base resistance, etc.) the two diodes 1, 2 exhibit noise as shown in section 5.2.3.

Furthermore, partition noise occurs, and the corresponding noise source given in section G.3 has to be implemented in parallel with the current source I_α.

As Fukui has shown (e.g. 1981), an estimate of the minimum noise figure achievable is given by

$$F_{\text{min}} \approx 1 + h\left\{1 + \left(1 + \frac{2}{h}\right)^{1/2}\right\} \tag{5.90}$$

where

$$h = \frac{I_0 R_b}{V_T}\left(\frac{f}{f_t}\right)^2 \tag{5.91}$$

As for other active devices, there exists a bias current $I_C(F_{\text{opt}})$ where F becomes a minimum.

5.4 FIELD-EFFECT TRANSISTORS

Because field-effect transistors are majority-carrier devices, no diffusion capacitance exists. Therefore extremely high cutoff frequencies are achievable. Because they are mainly planar devices where the conducting channel is located directed under the crystal surface the mobility is reduced with respect to the bulk mobility, usually to about one third of it. Only by avoiding surface scattering can higher values become possible, e.g. in the case of a buried channel FET (BFET) (Dekkers *et al.*, 1981); an example is given later in this chapter (Fig. 5.17).

Even higher mobilities than the bulk mobility are achievable by suppressing the Coulomb scattering in the active channel. This can be done by separating the doping atoms from the channel by using doped wide gap material along the (undoped) active channel (modulation doping). Especially at low temperatures ($\approx$77 K) extremely high mobilities become possible. If strained epitaxial layers are used, e.g. $Ga_{0.8}In_{0.2}As$ instead of GaAs (pseudomorphic FET) further improvement is achievable because of an even better suited band structure (e.g. Chao, *et al.*, 1989). However the principal electronic behaviour of those devices and the equivalent circuit correspond to conventional MESFETs, but higher g_m values and better RF and noise performance are achievable.

5.4.1 Electronic parameters

In this section the principal behaviour of FETs is discussed for GHz applications. The planar configuration requires characteristic parameters normalized to the length of the device perpendicular to the current flow, therefore, e.g. characteristic contact resistances ρ_c or the transconductance g_m are given in $\Omega\,\text{mm}^{-1}$, $\text{mS}\,\text{mm}^{-1}$, respectively. Thus devices with different geometries can be compared.

On the other hand the transit time t_t is in principle not dependent on the width but on the length of the active part of the device in the direction of the current flow. Therefore the length of the conducting channel underneath the gate stripe (width W) plays an important role in FET behaviour. Depending on the length of the gate area, velocity overshoot effects should occur if the gate length becomes comparable to the mean free path of the carriers (less than about 0.1 μm). This effect should lead to an enhancement of the carrier velocity along the channel. In practice, however, this has not clearly been seen so far.

To understand the upper frequency limit of the signal transfer in an FET structure the following assumption can be made. Strongly reduced control of the

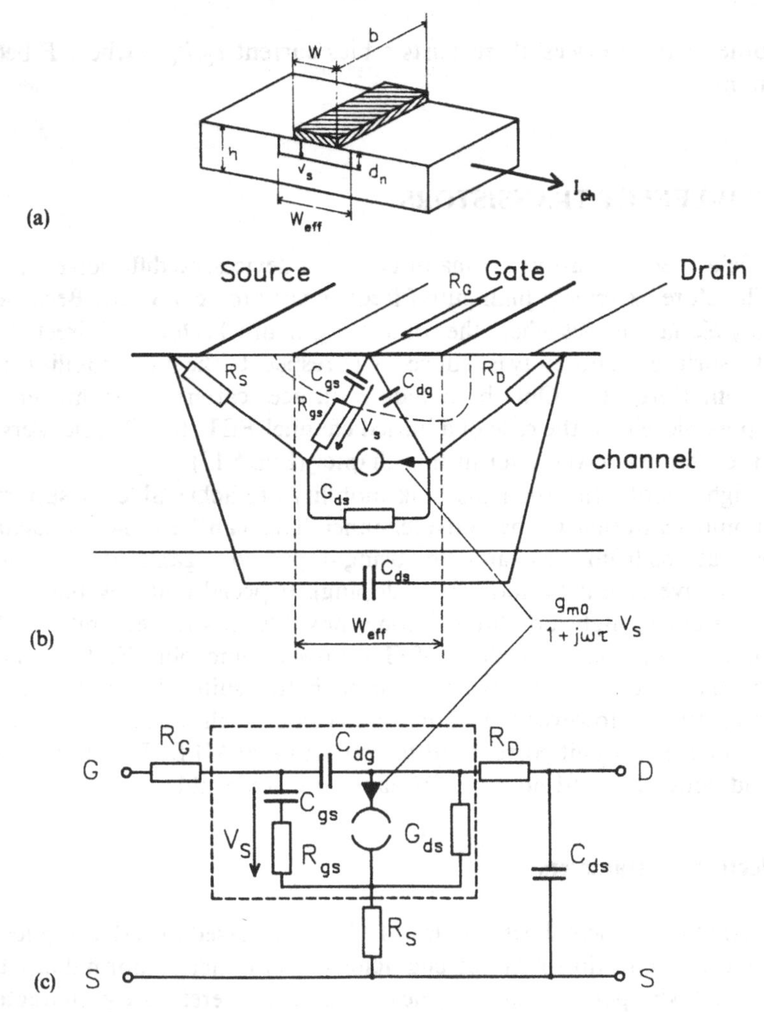

Fig. 5.16 Derivation of the FET equivalent circuit. (a) Physical model (JFET/MESFET); (b) corresponding lumped element structure; (c) equivalent circuit.

gate voltage on the carrier flow and therefore loss of gain will occur if the applied sinusoidal signal on the gate electrode changes its sign, while the induced current pulse has not yet left the gate region, length $W_{eff} > W$ (Fig. 5.16(a)). The corresponding transit time is

$$t_t = \frac{W_{eff}}{\bar{v}} \tag{5.92}$$

which at this critical frequency f_{cr} should be about equal to half the period of the applied signal.

Therefore

$$\omega_{cr} t_t \approx \pi \tag{5.93}$$

or

$$f_{cr} \approx \frac{1}{2t_t} = \frac{\bar{v}}{2W_{eff}} \tag{5.94}$$

This value is approximately equal to equation (5.41)

$$f_c = \frac{0.44}{t_t}$$

derived in section 5.1.2 as the cutoff frequency of pulse transfer for a given distance. This is also an approximation for f_{max}, the maximum frequency of oscillation. It is about three times the transit frequency defined in section 5.1.2, which is therefore again a figure of merit for an FET. As shown below, f_{max} can be higher than f_t in practical devices and indeed can reach $f_{cr} \approx 3f_t$, dependent on the device parameters.

The frequency $f_1 = f_t$ defines the frequency where the short-circuit current gain drops to unity, as shown in section 5.1.1. There a strongly simplified equivalent circuit for an FET was used (Fig. 5.1(a)). A more exact description allows the configuration in Fig. 5.16(c), where in the resulting equivalent circuit the inner transconductance g_m includes a phase delay by setting

$$g_m = g_{mo} e^{-j\omega\tau} \tag{5.95}$$

or, for $\omega\tau \ll 1$,

$$g_m = \frac{g_{mo}}{1 + j\omega\tau} \tag{5.96}$$

The delay time $\tau \approx t_t/2$ represents the inner transit effects.

As for all electronic devices, the equivalent circuit in its principal configuration

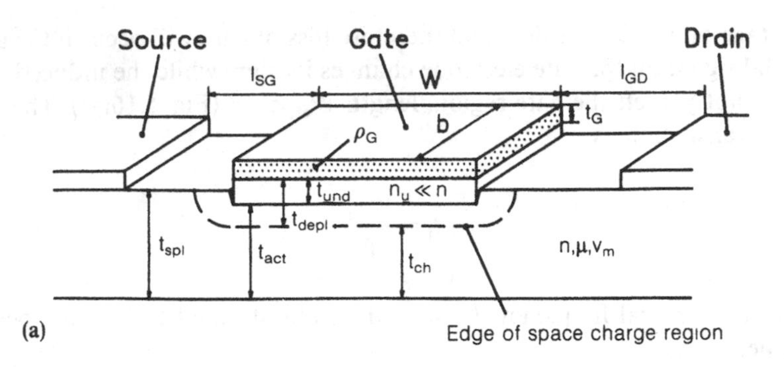

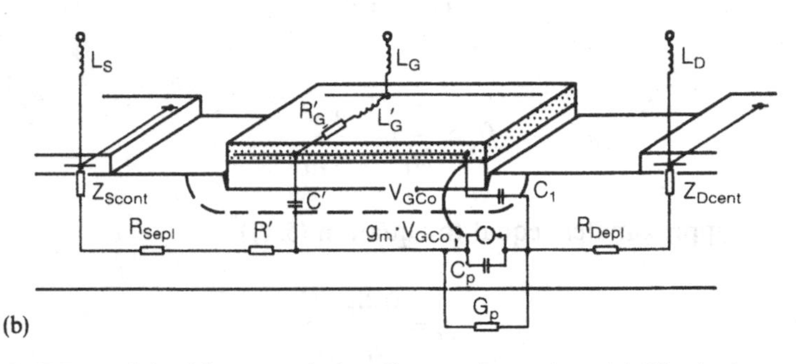

Fig. 5.17 BFET model with transmission line configuration. (a) Physical structure; (b) Equivalent circuit. (After Dekkers *et al.*, 1981.)

can be derived by inspecting the underlying geometrical configuration in conjunction with the physical behaviour of the device. This is demonstrated in Fig. 5.16(a), (b) as well as in Fig. 5.17. In the latter case a lightly doped intermediate layer underneath the gate electrode allows for higher linearity in the transfer characteristics (Fig. 6.3). Correspondingly the *S*-parameters of these buried channel FETs show minor frequency dependence as those of conventional field effect transistors allowing for better broadband matching (Dekkers *et al.*, 1981).

Taking into account the static domain in the channel region between gate and drain, developed by a dipole layer because of the differential negative mobility of electrons in III–V materials, a further capacitance C_{dc} has to be assumed between drain and the channel underneath the gate pad (Trew and Steer, 1987). This is shown in Fig. 5.18(a) for a MODFET, where in contrast to Fig. 5.16 the inner gate voltage is assumed to be applied along C_{gs} alone. The corresponding parameters are summarized in Table 5.1.

The resulting f_{max} values ($U = 1$) are modified (Table 5.2), and also the slope of the gain is modified. Instead of 6 dB per octave (20 dB per decade) a decay of 12 dB per octave (40 dB per decade) occurs at higher frequencies (Fig. 5.18).

If the gate width perpendicular to the current flow is relatively large the gate

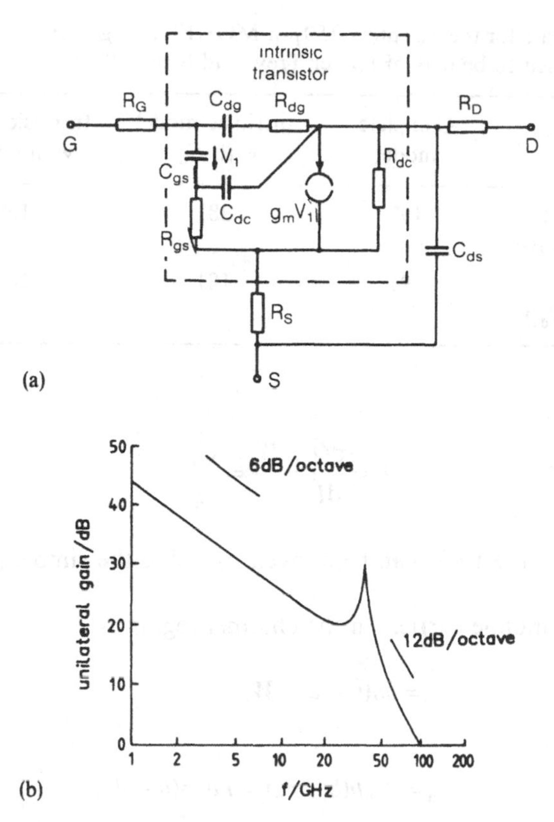

Fig. 5.18 Modified equivalent circuit of a FET (a) Lumped element model ($g_m = g_{m0}e^{-j\omega\tau}$); (b) gain behaviour. (After Trew and Steer, 1987.)

Table 5.1 Data for components of the equivalent circuit in Fig. 5.18(a) (after Trew and Steer, 1987)

$R_S =$	3.5 Ω	$C_{gs} =$	0.27 pF
$R_G =$	0.91 Ω	$C_{dg} =$	0.015 pF
$R_D =$	6.4 Ω	$C_{ds} =$	0.057 pF
$R_{gs} =$	4.62 Ω	$C_{dc} =$	10 fF
$R_{ds} =$	208.3 Ω	$g_{m0} =$	85.7 mS
$R_{gd} =$	2.8 Ω	$\tau =$	1 ps

pad has to be treated as a lossy transmission line, as shown in the geometrical configuration in Fig. 5.17(b) (Dekkers *et al.*, 1981). This reduces the electrically active gate width and enforces the use of multifinger arrangements in the upper GHz range.

That the general formulation of the transit time definition (section 5.1.2),

Table 5.2 Data for the 0.25 μm × 150 μm MODFET (Fig. 5.18) (the value of C_{dc} is assumed to be 0.01 pF) (after Trew and Steer, 1987)

f_{max}(GHz)	Complete model	Intrinsic model with C_{dc}	Intrinsic model without C_{dc}
Extrapolated 6 dB per octave	197	282	169
Calculated (including C_{dc})	91	181	160

$$t_t = \frac{d\bar{Q}}{d\bar{I}} = \frac{W}{\bar{v}} = \frac{C_i}{g_m} \tag{5.97}$$

is verified also for the FET, can be derived by using the simple physical model of Fig. 5.16(a).

The charge of mobile carriers in the channel region is

$$Q_{ch} = qn(h - d_n)bW_{eff} \tag{5.98}$$

and the current is

$$I_{ch} = J_{ch}b(h - d_n) = nq\bar{v}b(h - d_n) \tag{5.99}$$

Therefore

$$\frac{dQ_{ch}}{dI_{ch}} = \frac{Q_{ch}}{I_{ch}} = \frac{W_{eff}}{\bar{v}} \tag{5.100}$$

The gate-channel capacitance is

$$C_{ch} = \frac{\varepsilon W_{eff} b}{d_n} = \varepsilon W_{eff} b \left(\frac{qN_D}{2\varepsilon}\right)^{1/2} \frac{1}{(-(V_D + V_S))^{1/2}} \tag{5.101}$$

where $N_D = n$, V_D is the built-in potential and V_S the inner gate channel applied voltage and the transconductance is

$$g_m = \frac{dI_{ch}}{dV_S} = -bN_D q\bar{v}\frac{d(d_n)}{dV_S} \tag{5.102}$$

with

$$d_n = \left(\frac{2\varepsilon}{qN_D}\right)^{1/2} (-(V_D + V_S))^{1/2} \tag{5.103}$$

It also follows that

$$\frac{g_m}{C_{ch}} = \frac{\bar{v}}{W_{eff}} \tag{5.104}$$

which is equal to

$$\frac{1}{t_t} = 2\pi f_t = 2\pi f_1 \tag{5.105}$$

Neglecting the phase delay of the transconductance ($\tau = 0$), the equivalent circuit shown in Fig. 5.16(c) enables us also to calculate the power gain of the FET. With the parameters of Fig. 5.16(c) it follows, from the expression of the unilateral gain, that

$$U = \frac{|y_{21} - y_{12}|^2}{4(g_{11}g_{22} - g_{12}g_{21})} \tag{5.106}$$

(Appendix C) after deriving the y-parameters in the high frequency limit

$$U \approx \left(\frac{f_{max}}{f}\right)^2 \tag{5.107}$$

with the maximum frequency of oscillation ($R_D = 0$)

$$f_{max} \approx \frac{1}{4\pi}\left\{\frac{C_{dg}}{g_m} R_G C_{gs} + \frac{C_{gs}^2}{g_m^2} G_{ds}\left(R_{gs} + \frac{R_G + R_S}{1 + R_S g_m}\right)\right\}^{-1/2} \tag{5.108}$$

The inner device therefore exhibits ($R_G = R_S = 0$) the characteristic quantity

$$f_{max}|_{R_G = R_S = R_D = 0} = f_{maxi} = \frac{f_t}{2(R_{gs}G_{ds})^{1/2}} \tag{5.109}$$

leading to $(R_{gs}G_{ds})^{1/2}$ as a figure of merit in addition to f_t. This demonstrates the importance of a high output resistance $1/G_{ds}$ of a FET, responsible for high voltage gain. For $R_{gs}G_{ds} < 0.25$ it is $f_{maxi} > f_t$ as mentioned earlier in this section.

5.4.2 Noise parameters

The noise sources involved can be added to the equivalent circuit shown in Fig. 5.16(c), to derive a noise equivalent circuit. The feedback pad (C_{dg}) leads to complicated expressions. However, a semitheoretical evaluation leads to a simple formula for the minimum noise figure, first given by Fukui (e.g. 1981)

$$F_{\min} = 1 + K\frac{f}{f_t}(g_m(R_G + R_S)^{1/2} \quad (5.110)$$

where $K \approx 2.5$ $(1 \cdots 3)$ is a fitting factor depending on the channel material and its interface behaviour. Because

$$f_t = \frac{g_m}{2\pi C_{gs}} \quad (5.111)$$

it can also be written

$$F_{\min} = 1 + K2\pi f C_{gs}\left(\frac{R_G + R_S}{g_m}\right)^{1/2} \quad (5.112)$$

which shows the need for high g_m and small C_{gs} values, to achieve good noise performance. It should be mentioned that in general MODFETs exhibit somewhat lower $F_{\min}$ values than conventional MESFETs.

5.5 NEGATIVE RESISTANCES

Two groups of devices and circuits exist which exhibit negative impedance or negative admittance behaviour without any control mechanism from outside.

The first group are oneport (two-terminal) devices with an inner positive feedback mechanism resulting in piecewise negative DC characteristics (Fig. 5.19). The second group are circuit configurations where at the entrance port connected

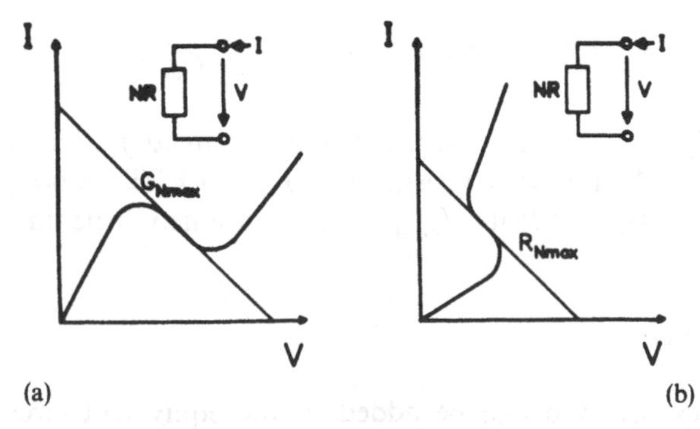

Fig. 5.19 Nonlinear DC characteristics with piecewise negative slope (NR device). (a) Negative conductance VC device; (b) negative resistance CC device.

conventional passive components (resistances, capacitances, inductances) become transformed to negative impedances or admittances at the output port. The negative impedance converter (NIC) used to reduce attenuation and distortion of telephone channels is an example of the second group, whereas Shockley diodes, tunnel diodes or resonant tunnel barriers belong to the first group, applicable to RF amplification or switching.

The parametric amplifier (paramp) where de-attenuation is achieved by parametric variation of reactances is another configuration to achieve negative resistances. In this case a pump source acts as a control mechanism to modulate the stored energy in the reactance. Paramps allow extremely low noise (narrowband) RF amplification. A short description is given in this chapter (also section E.2.2 and Exercise 5.4).

5.5.1 Oneport devices

If properly biased the oneport devices with piecewise negative characteristics allow the construction of fast switching or memory (bistable) circuits and can be used for amplification of high frequency signals or as mixers and oscillators (e.g. the tunnel diode, sections 5.2.2 and E.2.1).

(a) DC characteristics and biasing

We distinguish here between voltage-controlled (VC) and current-controlled (CC) negative resistance (NR) devices. In the first case (Fig. 5.19(a)), the current I through the device is an unequivocal function of the applied voltage V, but not

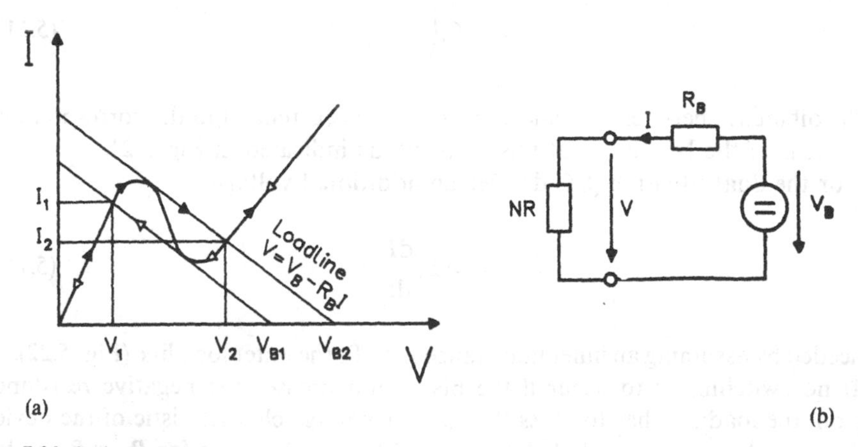

Fig. 5.20 Switching behaviour of a voltage-controlled device. Indicated is the switching path if V_B crosses the critical value V_{B2} (for increasing V_B) and V_{B1} (for decreasing V_B), respectively. (a) Graphical representation; (b) circuit configuration.

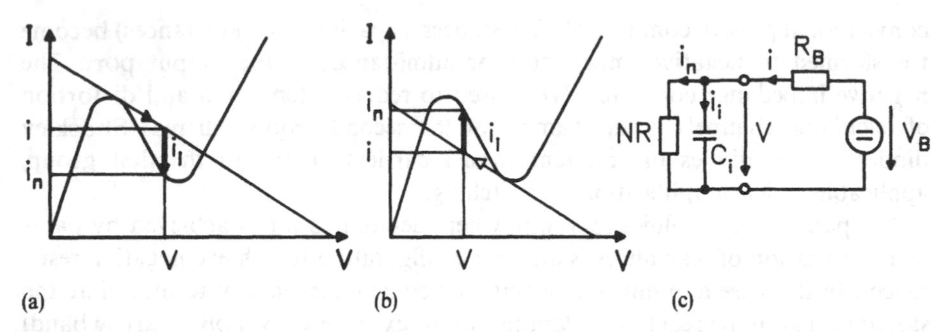

Fig. 5.21 The switching process in a VC device. (a) Turn-on; (b) turn-off; (c) circuit representation.

vice versa. CC devices exhibit dual behaviour, as can be concluded from Fig. 5.19(b).

In Fig. 5.20(b) a simple biasing circuit is shown, where the bias point along the static characteristic according to Fig. 5.20(a) moves suddenly over the negative resistance regime, if the applied voltage exceeds a critical value. Solid arrows indicate the movement of the bias point if the applied voltage grows, and open arrows indicate the reverse switching path. Dual behaviour occurs in the case of a CC device. As a result, hysteresis occurs, as shown in Fig. 5.20(a) for the VC device. The stable endpoints are $\{V_1, I_1\}$ for the low state (off) and $\{V_2, I_2\}$ for the high state (on).

Along the switching cycle, the DC load line (R_B) does not match the steady state characteristic of the NR device at any time. To overcome this problem, an inner capacitance C_i is assumed, where the induced current

$$i_i = C_i\left(\frac{dV}{dt}\right) \tag{5.113}$$

is the difference between the time-dependent bias current i and the corresponding current i_n of the NR device at this moment, as indicated in Fig. 5.21.

For the dual situation (CC device), an additional voltage

$$v_i = L_i\frac{dI}{dt} \tag{5.114}$$

is needed by assuming an inner inductance L_i to fit the outer load line (Fig. 5.22).

If no switching is to occur if the bias point crosses the negative resistance region, the load line has to cross the current-voltage characteristic of the device only once. As can be concluded from Fig. 5.19, this happens for $R_B < R_{Nmax}$ for the VC device and for $R_B > R_{Nmax}$ for the CC device, R_{Nmax} being the steepest slope of the DC characteristic in the negative resistance region. The more

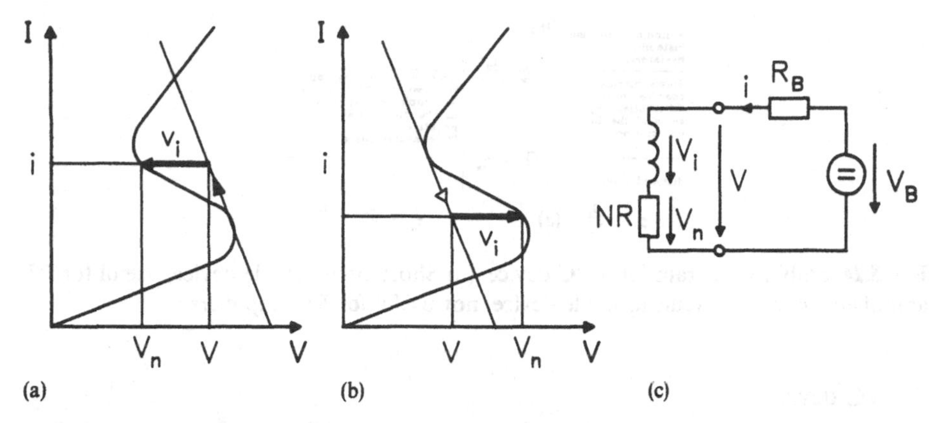

Fig. 5.22 The switching process in a CC device. (a) Turn-on; (b) turn-off; (c) circuit representation.

consistent and therefore better description however, for the VC device is to express the above given relationship by $G_B > G_{Nmax}$, using the conductances $G_B = 1/R_B$ and $G_N = 1/R_N$. This indicates more clearly that for the VC device to become stable the sum of the conductances has to be positive. Consequently the CC device is a **negative resistance**, and the VC-device a **negative conductance** (next section).

However, the conditions given above are not sufficient for absolute stability. In Fig. 5.23 the equivalent circuits of both types are given, including the parasitic components L_s (series inductance), R_s (series resistance) and parasitic capacitance C_p. The LC combination acts as a transformer at a given frequency, leading to a second condition for stability.

The outer biasing circuit may be connected in series or parallel to the applied RF circuitry, leading to a similar equivalent circuit (Fig. 5.21(c)). With $R_s + R_B = R$, $L_s + L_L = L$, $C_i + C_p = C$ it follows in both cases for the resulting conductance G_t parallel to $-G_N$, as long as C_B is sufficiently small and is neglected, that for

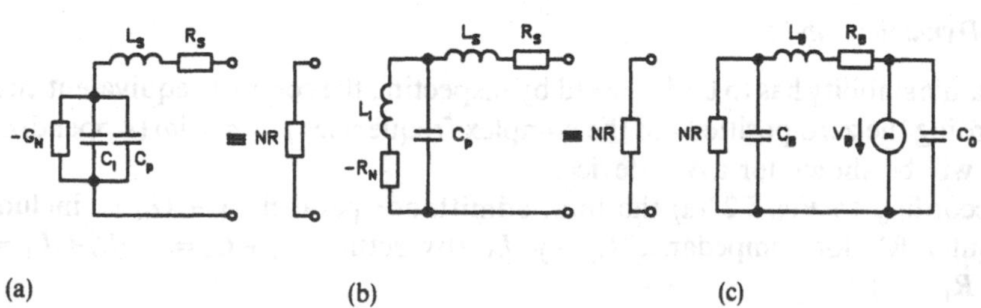

Fig. 5.23 Equivalent circuits of NR devices in the negative resistance regime. (a) VC device; (b) CC device; (c) complete circuit with bias voltage and shunting capacitance ($C_0 \rightarrow \infty$).

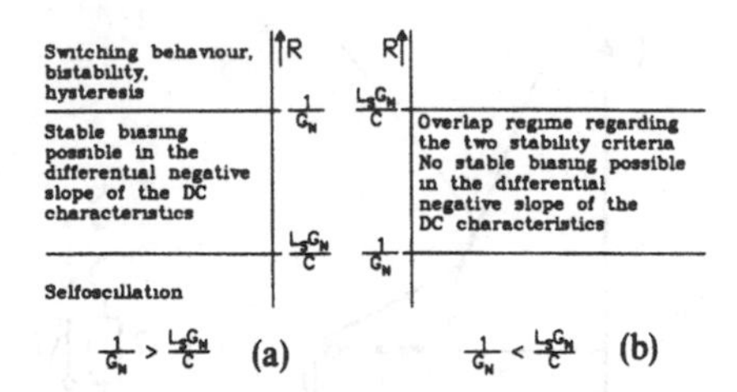

Fig. 5.24 Stability diagram for a VC device. (a) Short-circuit stable device, useful for RF amplifiers; (b) Short-circuit unstable device, not useful for RF amplifiers.

the VC device

$$G_t = \frac{CR}{L} \tag{5.115}$$

Therefore to achieve DC stability the whole resistance R is subject to the condition

$$\frac{1}{G_N} > R > \frac{LG_N}{C} \tag{5.116}$$

If this relationship is not fulfilled for the NR device itself, the device can never be biased in a stable way; therefore the VC device can only be biased stably if the inequality

$$\frac{1}{G_N} > R_s > \frac{L_s G_N}{C} \tag{5.117}$$

is valid.

The graphical representation in Fig. 5.24 shows the different regions. As long as stability exists, the device can be used in the negative dynamic impedance regime for RF amplification, oscillator application, etc.

(b) Dynamic stability

Dynamic stability has to be discussed by inspecting the complete equivalent circuit regarding the zero positions of the complex frequencies $p = \sigma + j\omega$ (Appendix D). This will be shown for a VC device.

According to Fig. 5.23(a) the total admittance parallel to $-G_N$ is, including an outer RF load impedance $R_L + j\omega L_L$ (by setting $C_i + C_s = C$, $L_s + L_L = L$, $R_s + R_L = R$),

$$Y = -G_N + j\omega C + (j\omega L + R)^{-1} \tag{5.118}$$

Introducing p for jω the roots for $Y(p)=0$ can be determined. They are $p_{1,2}=\sigma_0 \pm \mathrm{j}\omega_0$ with

$$\sigma_0 = \frac{G_N L - RC}{2LC} \tag{5.119}$$

and

$$\omega_0 = \left(\frac{1-RG_N}{LC} - \left(\frac{RC - G_N L}{2LC} \right)^2 \right)^{1/2} \tag{5.120}$$

Turn-on behaviour and therefore instability would occur for $G_N L > RC$ as concluded in section 5.5.1(a).

Two types of admittance curves in the complex frequency plane of the NR device itself are possible. They are drawn in Fig. 5.25 together with the ideal behaviour

$$Y_N = -G_N + \mathrm{j}\omega C_i \tag{5.121}$$

without parasitics. Both curves cross both axes for growing frequency but rotate in a different sense. They exhibit two characteristic frequencies, where either the real part or the imaginary part of the admittance becomes zero (Fig. 5.25). These are the attenuation corner frequency

$$f_a = f|_{\mathrm{Re}(Y)} = 0 \tag{5.122}$$

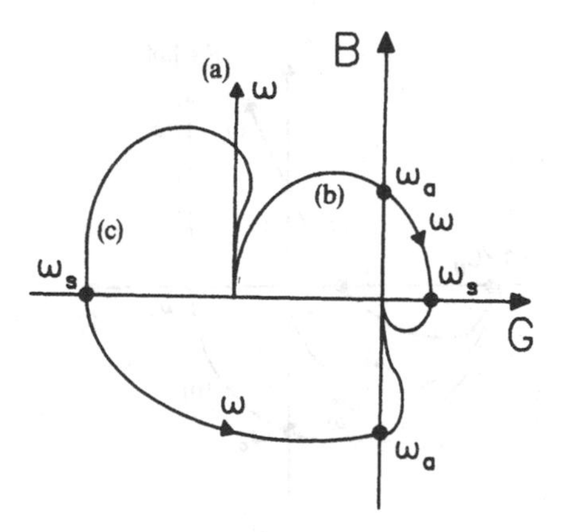

Fig. 5.25 Admittance curves $G+\mathrm{j}B$ in the complex $\{G, B\}$ plane for a VC device in the negative conductance regime. (a) Ideal configuration without parasitic components; (b) admittance curve for $\omega_s > \omega_a$ (short-circuit stable); (c) admittance curve for $\omega_s < \omega_a$ (short-circuit unstable).

and the self-resonant frequency

$$f_s = f|_{\mathrm{Im}(Y)} = 0 \tag{5.123}$$

The corresponding ω-values follow with

$$\omega_a = \frac{R}{L}\left(\frac{1}{RG_N} - 1\right)^{1/2} \tag{5.124}$$

and

$$\omega_s = \frac{R}{L}\left(\frac{L}{R^2C} - 1\right)^{1/2} \tag{5.125}$$

If $\omega_a < \omega_s$, then no self-oscillation is possible, because at the critical frequency f_s no deattenuation occurs; the VC device under consideration is short-circuit stable. On the other hand, if $\omega_a > \omega_s$, then instability occurs. The limit is $\omega_a = \omega_s$, which corresponds to the above given relationship

$$\frac{1}{G_N} = \frac{L}{RC} \tag{5.126}$$

The same can be concluded by inspection of the corresponding impedance curves (Fig. 5.26).

Similar conclusions can be drawn for a CC device by inspecting the dual equivalent circuit (Fig. 5.23(b)).

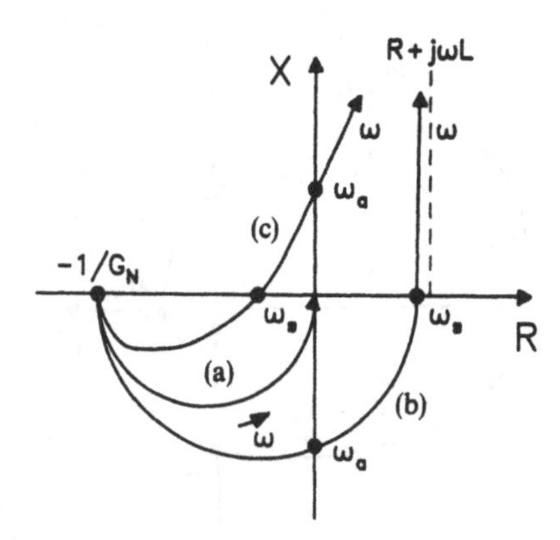

Fig. 5.26 Impedance curves $R + jX$ in the complex $\{R, X\}$ plane for a VC device in the negative conductance regime. (a) ideal configuration without parasitic components; (b) admittance curve for $\omega_s > \omega_a$ (short-circuit stable); (c) admittance curve for $\omega_s < \omega_a$ (short-circuit unstable).

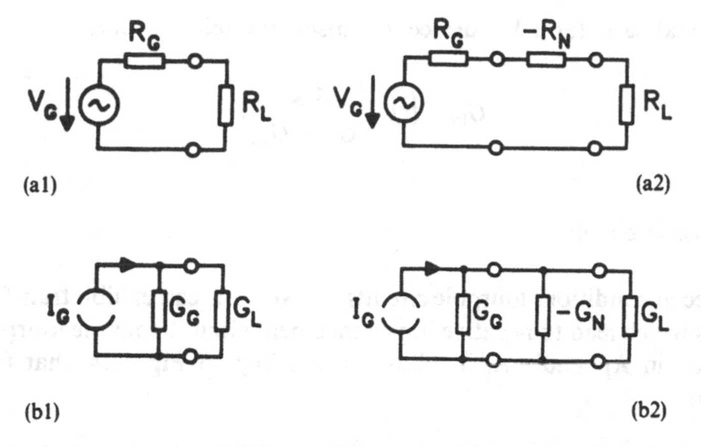

Fig. 5.27 Equivalent circuits of NR amplifiers (parasitics neglected). (a1) Conventional generator–load connection; (a2) NR amplifier with CC device; (b1) Conventional generator–load connection; (b2) NR amplifier with VC device.

(c) Amplification

To evaluate the differential gain G of an amplifier with negative resistance or conductance a conventional generator–load circuit and the amplifier with the NR device have to be compared (Fig. 5.27).

At high frequencies reflection amplifiers instead of the conventional insertion amplifier are used, where the input power and output power are separated by circulators (section E.1).

In the conventional configuration, shown simplified without any parasitic components in Fig. 5.27, the power in the load is, in the case of a CC device

$$P_L = \frac{V_G^2 R_L}{(R_G - R_N + R_L)^2} \tag{5.127}$$

which has to be compared with

$$P_{LO} = \frac{V_G^2 R_L}{(R_G + R_L)^2} \tag{5.128}$$

The insertion gain is the ratio of both,

$$G_{IN} = \left(1 - \frac{R_N}{R_G + R_L}\right)^2 \tag{5.129}$$

Therefore the reduction of the total ohmic resistance of the circuit by the negative resistance R_N is responsible for the gain, leading to instability for $R_N = R_G + R_L$.

In the dual case, for a VC device, the insertion gain becomes

$$G_{\mathrm{IN}} = \left(1 - \frac{G_{\mathrm{N}}}{G_{\mathrm{G}} + G_{\mathrm{L}}}\right)^2 \tag{5.130}$$

5.5.2 Fourpole circuits

Under special conditions fourpole circuits or twoports can exhibit transformation properties which lead to negative impedance behaviour. From the fourpole equations, given in Appendix A, it follows, according to Fig. 5.28, that for the *h*-parameters

$$Y_2 = h_{22} - \frac{h_{12}h_{21}}{h_{11} + Z_{\mathrm{G}}} \tag{5.131}$$

and for the *p*-parameters

$$Z_2 = p_{22} - \frac{p_{12}p_{21}}{p_{11} + Y_{\mathrm{G}}} \tag{5.132}$$

If the parameters h_{22}, h_{11} and p_{22}, p_{11}, respectively, are sufficiently small, then the connected input immitances Z_{G} and Y_{G} are transformed to negative immitances

$$Y_2 \approx -(h_{12}h_{21})Y_{\mathrm{G}} \tag{5.133}$$

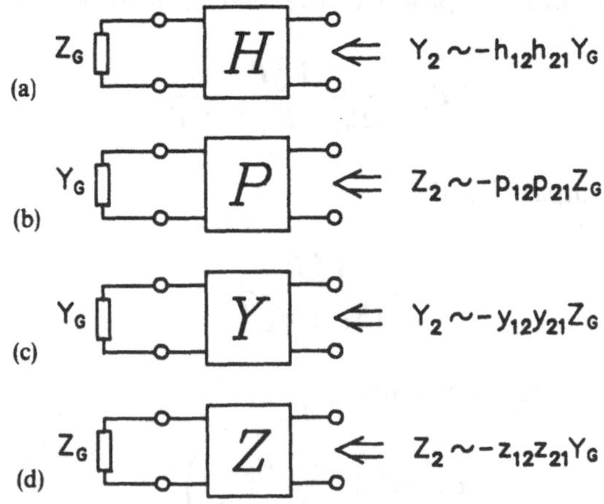

Fig. 5.28 Principal configurations of impedance and admittance converters (limits given in text). (a) Admittance → (neg.) admittance; (b) impedance → (neg.) impedance; (c) admittance → (neg.) inverted impedance; (d) impedance → (neg.) inverted admittance.

and

$$Z_2 \approx -(p_{12}p_{21})Z_G \tag{5.134}$$

as long as the transformation constants $(h_{12}h_{21})$ and $(p_{12}p_{21})$ are positive.

Similar results can be achieved via the y- and z-parameters, where additionally an inversion occurs; the output exhibits the inverse frequency behaviour to the connected immitance at the input; e.g. a capacitance connected to the input becomes transformed to a negative inductance at the output.

In section E.1 the use of an operational amplifier for the construction of a negative impedance converter (NIC) is shown.

5.5.3 Parametric amplifiers

In the case of conventional amplifying devices the gain mechanism is based on variable ohmic components in the active structure. However, varying reactances L, C also allow RF amplification at a signal frequency f_s because of their energy-storage behaviour, if special conditions are fulfilled (Manley and Rowe, 1957). For the continuous dynamic variation of the reactances an RF pump generator, with frequency f_p, is needed. (A mechanical analogy is the swing, where special phase relationships are afforded between the excitation movement and the resonant frequency of the swing.) By using a third resonance frequency f_i in the circuit of the paramp, $f_i = f_p - f_s$ (f_i is the idler frequency, Fig. 5.29), the impeding phase relationship can be skipped.

Paramps are of practical interest at very high frequencies where conventional active devices, e.g. MESFETs, are not applicable. They exhibit very good noise performance because only parasitic components contribute to their thermal noise. If the system is cooled, excellent noise figures are achievable (Uenohara, 1960) (Appendix E.2.2).

For paramps the use of varactor diodes is advantageous because of their large voltage-dependent capacitance variations and good RF performance. The alternating part of the capacitance is a nonlinear function of the applied (reverse) voltage. However, for a simplified analysis, linearization is acceptable:

$$C = C_0 + C_1 \cos(\omega_p t) \tag{5.135}$$

The modulation frequency (ω_p) is the sweep frequency of the activating pump signal, coming from a local oscillator.

If the varactor diode is connected in series, for the circuit analysis the inverse function $1/C$ has to be linearized, leading to

$$\frac{1}{C} = \frac{1}{C_0 + C_1 \cos(\omega_p t)} \approx \frac{1}{C_0} - \frac{C_1}{C_0^2}\cos(\omega_p t) \tag{5.136}$$

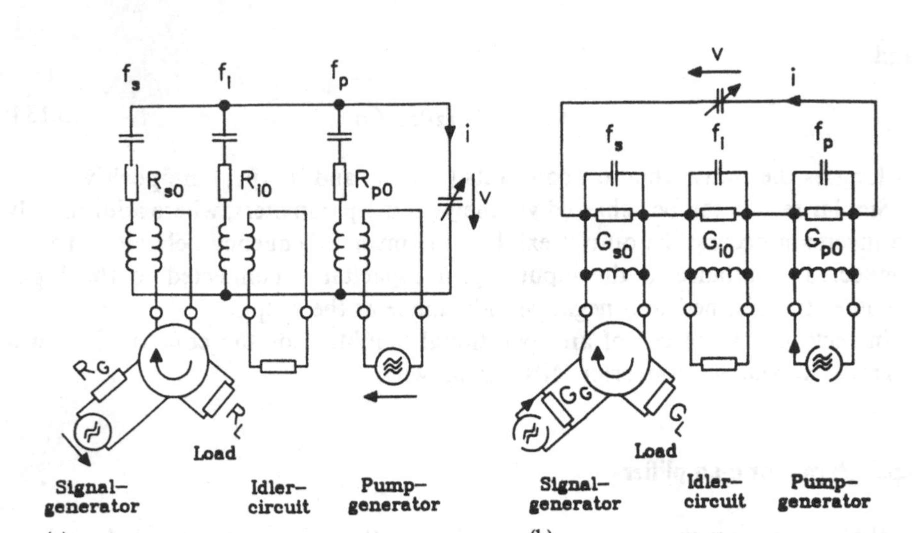

Fig. 5.29 Three-frequency parametric amplifier ($f_p = f_i + f_s$). Indicated is a circulator for the separation of input and output at the signal frequency (idealized). (a) Series resonant circuits; (b) parallel resonant circuits.

Whereas for the varying susceptance the quantity C_1 or C_1/C_0 is a figure of merit. In the case of the modulated reactance it is the elastance C_1/C_0^2.

With the stored charge $Q = CV$ in the junction which acts as the condenser with varying capacitance, the resulting current through the diode becomes

$$i = \frac{dQ}{dt} = \frac{d(CV)}{dt} = C\frac{dV}{dt} + V\frac{dC}{dt} \tag{5.137}$$

The second term leads to negative conductances or negative resistances, by proper choice of the circuit and the relevant frequencies.

As mentioned above, the most common arrangement is a three-frequency configuration with either three series resonant circuits or three parallel resonant circuits (Fig. 5.29). In these configurations only currents and voltages of the three frequencies, pump frequency f_p, idler frequency f_i and signal frequency f_s, can be excited. If the relationship $\omega_p = \omega_i + \omega_s$ is established and the pump generator activates the varactor diode, then at both frequencies f_i, f_s gain becomes possible corresponding to the Manley–Rowe relationship (Manley and Rowe, 1957)

$$\frac{P_p}{\omega_p} = \frac{P_s}{\omega_s} = \frac{P_i}{\omega_i} \tag{5.133}$$

where P_p is the power delivered by the local oscillator at f_p, and P_s and P_i are the resulting output powers at f_s and f_i, respectively.

As a result, negative impedances occur at f_i and f_s, leading at resonance to a negative conductance

$$-G_N = -\frac{\omega_s \omega_i C_1^2}{4G_{i0}} \tag{5.139}$$

parallel to the corresponding port in Fig. 5.29(b), and to a corresponding negative series resistance in Fig. 5.29(a) (G_{i0} is the residual ohmic conductance at f_i).

To separate the output signal from the incoming signal, practical reflection type paramps have to use circulators, as indicated in Fig. 5.29. For amplifiers inserted in the transmission line, directional couplers or unilines have to be used at input and output, otherwise stability problems occur (section E.2).

Besides amplification at the signal frequency, upconversion can be used. If f_i is chosen much higher than f_s, high gain can be achieved. From the Manley–Rowe relationship it follows that for $f_p \gg f_s$ the power at the idler frequency is

$$P_i = P_p \frac{\omega_p - \omega_s}{\omega_p} \tag{5.140}$$

or with

$$P_s = P_p(\omega_s/\omega_p),$$

$$P_i = P_s \frac{\omega_p - \omega_s}{\omega_s} \gg 1 \tag{5.141}$$

5.6 PHOTONIC DEVICES

As mentioned in the introduction, photonic devices are not treated separately. Many materials show inherently optical effects or electro- and magneto-optical interactions which can be used for optoelectronic devices. The effects are based on variations of the band structure and modified energy levels of holes and electrons in the material.

Regarding optoelectronic devices the transfer efficiency η is a characteristic quantity, e.g. the number of generated electron–hole pairs in relation to absorbed energy quanta (photons) $h\nu$ if detectors are discussed, or for emitting devices the number of generated light quanta per recombining electron–hole pair.

Out of the efficiency of the transfer process the interaction time is of interest, how fast the generation of electron–hole pairs occurs, or how fast the recombination energy is transformed into a photon. Excitation (energy gain of an electron or hole, heating up) or cooling down (energy loss by any kind of interaction, thermalization) are very fast processes, of the order of 10 fs. Only by participation of traps can the time scale be extended to ps, ms, or even hours. From an application point of view the transfer time itself is negligible in comparison with other time constants in an (opto-) electronic device, and only those must be taken into account. A photon is never stored as a quantum of light anywhere (an exception is the high-Q optical resonator). The energy is always stored on the electronic side.

EXERCISES

5.1 A photodiode has an illumination-dependent DC characteristic shown in Fig. Ex5.1(a). The equivalent circuit of the (p–n) diode consists, in the reverse direction, of a current source I_{ph}, which is dependent on the absorbed light, in parallel with the inner capacitance C_r, which is assumed to be constant (Fig. Ex5.1(b)). The dynamic inner conductance G_r and a

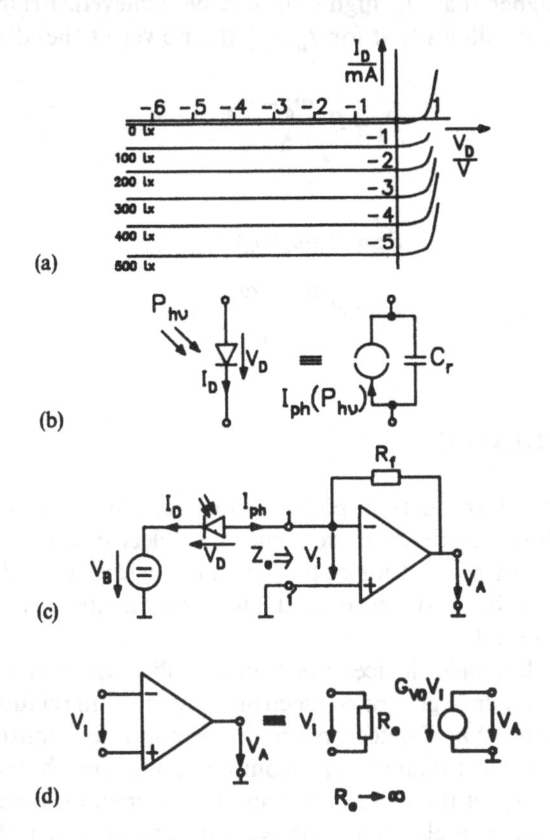

Fig. Ex5.1 Photoreceiver. (a) Photodiode characteristics; (b) photodiode equivalent circuit; (c) circuit with operational amplifier; (d) OP equivalent circuit.

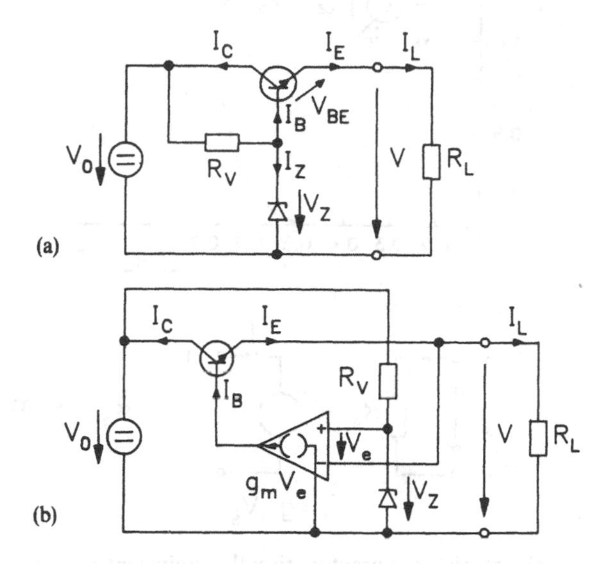

Fig. Ex5.2 Circuits for voltage stabilization: (a) without operational amplifier; (b) with operational amplifier.

reverse current I_0 without illumination are neglected (Fig. Ex5.1(b)). To achieve an illumination-dependent voltage instead of a current, an operational amplifier (OP) as a current–voltage converter is used, with $R_f = 1\,\text{M}\Omega$ (Fig. Ex5.1(c)). The bias voltage is $V_B = 5\,\text{V}$. Determine the current I_{ph} for an illumination of 500 lx and the resulting output voltage V_o. Assume an ideal OP with open circuit voltage gain $G_{V_0} \to \infty$.

Compute the input impedance Z_e of the circuit at point 1, 1′. If a capacitance of $C_r = 100\,\text{pF}$ is assumed, determine the cutoff frequency f_c of $G_V = V_A/V_i$; $G_V(f = f_c) = G_V(f = 0)/2^{1/2}$.

5.2 A voltage-stabilizing circuit is shown in Fig. Ex5.2. The Zener voltage V_Z across the Zener diode is assumed to be constant. If no OP is used (Fig. Ex5.2(a)), determine the resistance R_v for $I_Z = 100\,\text{mA}$, $V_0 = 10\,\text{V}$, $V_Z = 5\,\text{V}$ ($I_Z \gg I_B$). What are the voltage V and base current I_B for $I_L = 1\,\text{A}$?

The DC current gain is assumed to be $I_C/I_B = B = 100$, the input current $I_E(V_{BE})$ follows from the static characteristic given in Fig. Ex5.3(a).

Determine the inner resistance

$$R_i \approx -\left.\frac{\Delta V}{\Delta I_L}\right|_{I_L = 1\,\text{A}}$$

of the stabilizing circuit (Fig. Ex5.2(a)).

If the OP is implemented (Fig. Ex5.2(b)), a drastic reduction of the inner

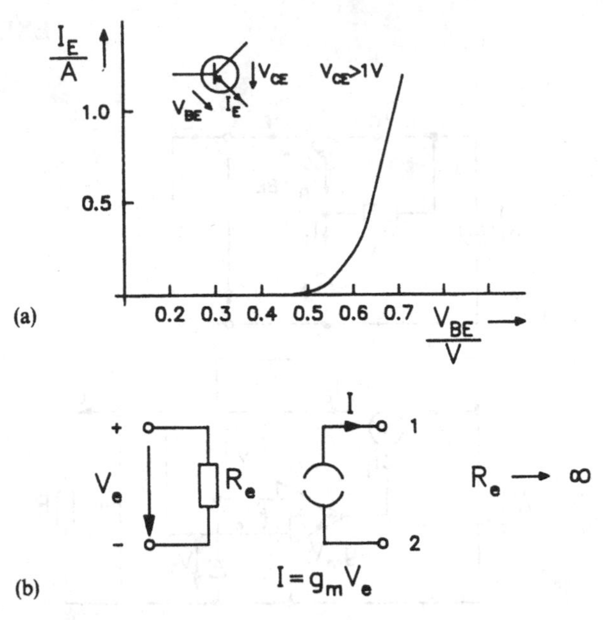

Fig. Ex5.3 (a) Transistor characteristics; (b) equivalent circuit of the OP.

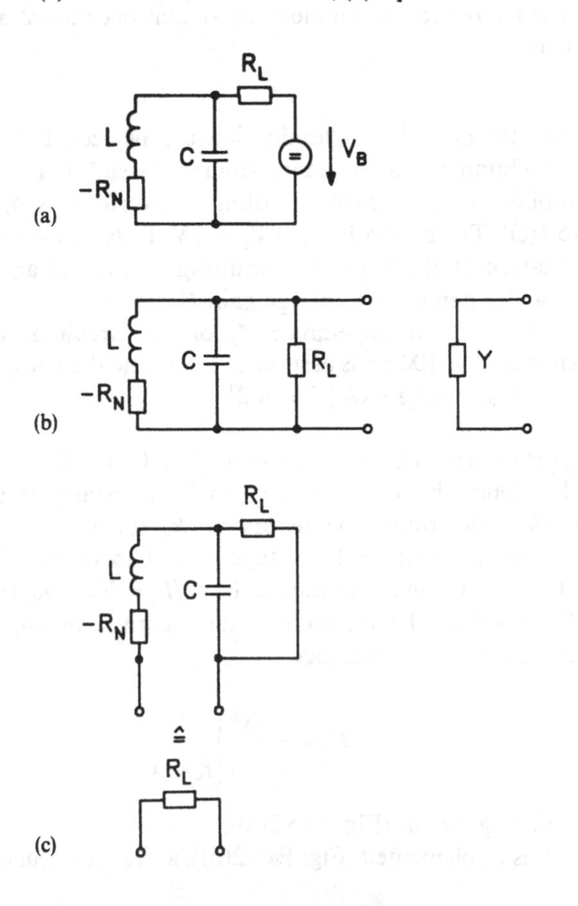

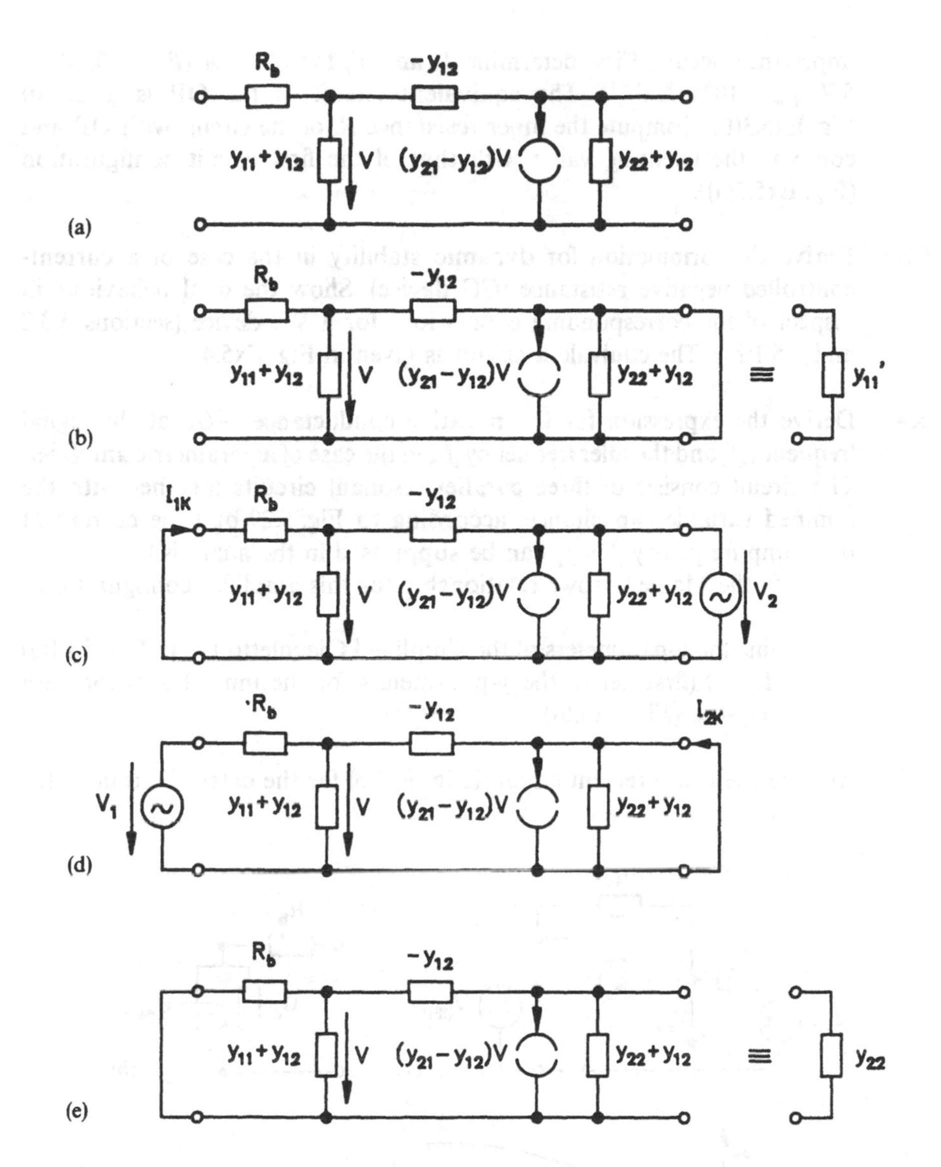

Fig. Ex5.5 *Y*-parameters of a bipolar transistor with base resistance (EB; *y*-parameter inner transistor, *y'*-parameter including R_b) (a) Equivalent circuit; (b) Determination of y'_{11}; (c) Determination of y'_{12}; (d) Determination of y'_{21}; (e) Determination of y'_{22}.

Fig. Ex5.4 Current-controlled negative resistance. (a) Equivalent circuit of a CC device in the negative impedance region (including the bias circuitry); (b) admittance connection; (c) impedance connection.

impedance occurs. First determine V and V_e for $I_L = 1$ A ($B = 100$, $V_Z = 5$ V, $g_m = 10^4$ mA V^{-1}). The equivalent circuit of the OP is given in Fig. Ex5.3(b). Compute the inner resistance R_i of the circuit with OP and compare the resulting value with that of the first circuit configuration (Fig. Ex5.2(a)).

5.3 Derive the formulation for dynamic stability in the case of a current-controlled negative resistance (CC device). Show the dual behaviour in respect of the corresponding expressions for a VC device (sections 5.3.2 and 5.5.1(b)). The equivalent circuit is given in Fig. Ex5.4.

5.4 Derive the expression for the negative conductance $-G_N$ at the signal frequency f_s and the idler frequency f_i, in the case of a parametric amplifier. The circuit consists of three parallel resonant circuits together with the pumped variable capacitance, according to Fig. 5.29(b). (The current at the pump frequency $f = f_p$ can be suppressed in the analysis.)

Verify the Manley–Rowe relationship for this amplifier configuration.

5.5 Determine the y-parameters of the simplified Giacoletto EC in Fig. 5.14(a) for $L_b, L_c = 0$ (first derive the y-parameters for the inner transistor with $R_b = 0$, $\omega_t \to \infty$ (Fig. Ex5.5)).

5.6 Analyse the measurement circuit (Fig. Ex5.6) for the determination of the

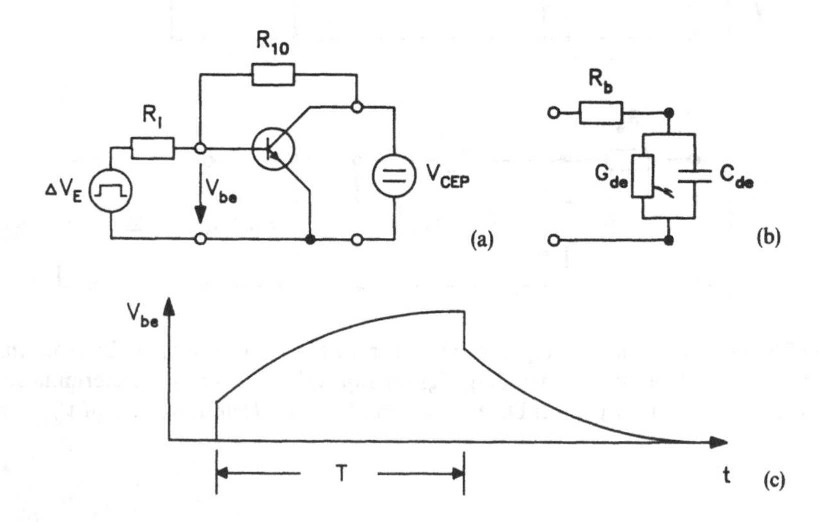

Fig. Ex5.6 Measurement circuit for R_b and diffusion time constant t_d of a bipolar transistor (small signal case). (a) Circuit with indication of the applied digital input signal (bias condition via $R_{10} \gg R_b$); (b) assumed equivalent circuit of the entrance port of the transistor (common emitter); (c) measured voltage response $V_{be}(t)$.

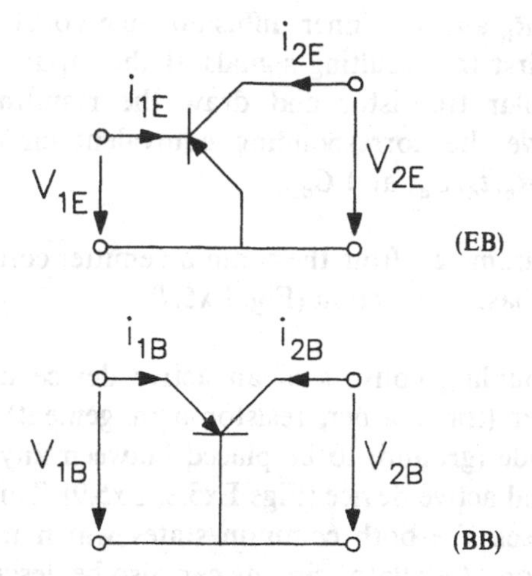

Fig. Ex5.7 Common emitter (EB) and common base (BB) configuration with indication of currents and voltages.

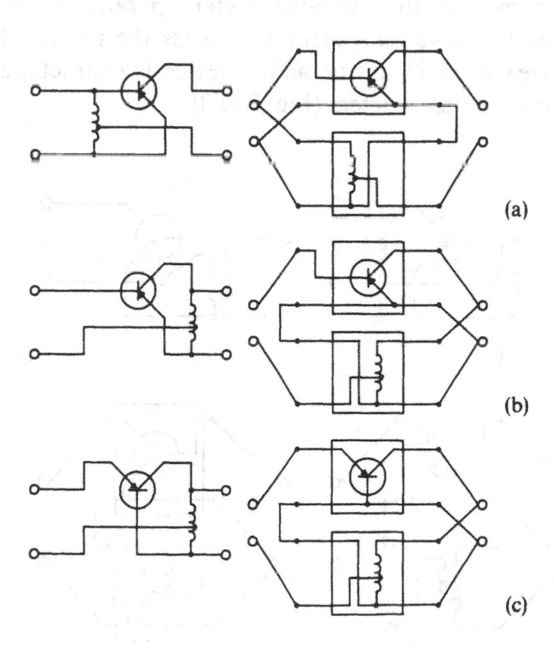

Fig. Ex5.8 Interbase connections, circuits and fourpole equivalent using bipolar transistors as the active device. (a) Common emitter/common base (EB/BB); (b) common emitter/common collector (EB/CB); (c) common base/common collector (BB/CB). (After Beneking, 1955.)

base resistance R_b and the inner diffusion time constant t_d of a bipolar device. Derive first the resulting signals at the input (ΔV_{BE}) and output (ΔI_C) of a bipolar transistor and draw the resulting time-dependent responses. Derive the corresponding equivalent circuits and give the expressions for R_b, t_d, G_{de} and C_{de}.

5.7 Convert the *h*-parameters from the common emitter configuration to those of the common base connection (Fig. Ex5.7).

5.8 An interbase coupling consists of an active device in connection with a voltage divider (transformer, resistor arrangement) which allows the common electrode (ground) to be placed between any two electrodes of the three-terminal active device (Figs Ex5.8, Ex5.9). This leads to a circuit behaviour between the both common states which might be useful for special applications. Oscillator circuits can also be described and analysed by interbase couplings (e.g. Exercise 4.5, Fig. Ex4.4).

In principle more than one transformer can be used, each between two electrodes of the three-terminal device, creating an unconventional fourpole with a wide span of parameters (Gupta, 1977).

Evaluate the fourpole parameters of the common emitter–common base combination by applying the *p*-parameters (parallel–series connection of the active and passive fourpole) and discuss the result. The transformer should be treated as ideal; the active device is characterized by its conventional fourpole parameters (Fig. Ex5.9).

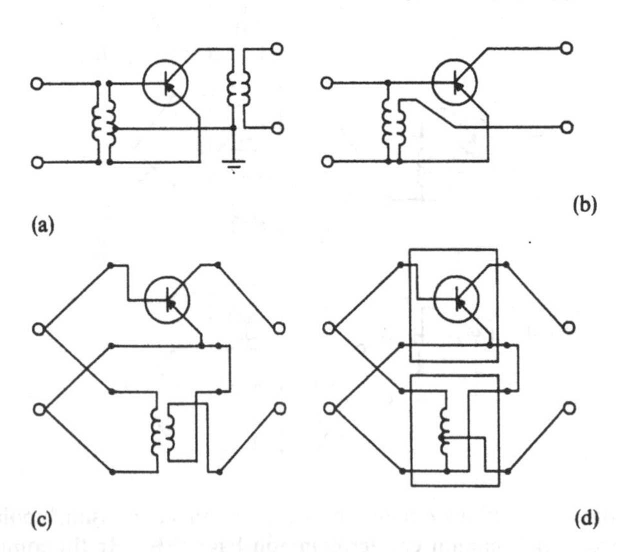

Fig. Ex5.9 Interbase coupling EB/BB. (a) RF circuit with matching transformers; (b) simplified circuit; (c) separation of the active and passive part; (d) *P*-matrix coupling.

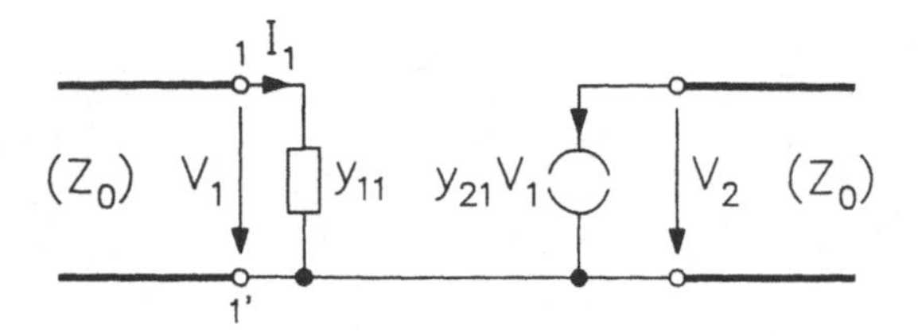

Fig. Ex5.10 Transistor with simplified EC in transmission line.

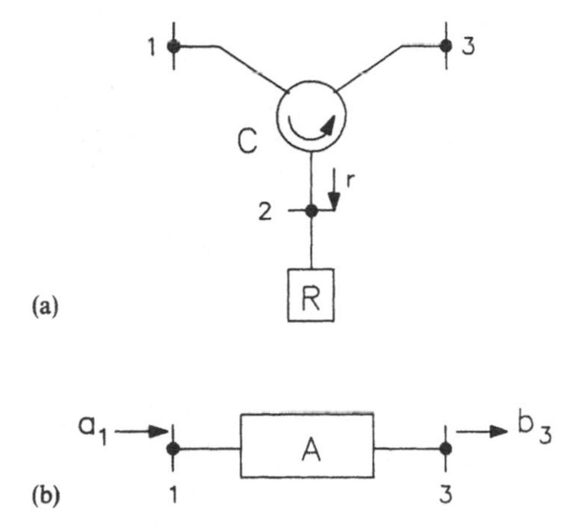

Fig. Ex5.11 Reflection amplifier with circulator. (a) Combination of the negative resistance device R with the circulator C; (b) resulting twoport amplifier A.

5.9 Determine the S-parameters of an idealized transistor according to Fig. Ex5.10 at the circuit planes 1–1′ and 2–2′.

5.10 The combination of a reflection amplifier R and a circulator C leads to a twoport amplifier (Fig. Ex5.11).

Derive the amplification of the resulting twoport amplifier in decibels, if R is characterized by its reflection factor

$$r = 10e^{j(30°)}$$

and C by its scattering matrix

$$\boldsymbol{S}_C = \begin{bmatrix} 0 & 0 & \tau \\ \tau & 0 & 0 \\ 0 & \tau & 0 \end{bmatrix}$$

with $\tau = 0.9$ ($e^{j(30°)}$ has to be read as $e^{j\pi/6}$).

Fig. Ex5.[illegible] Transistor with [illegible]

Fig. Ex5.[illegible] [illegible] negative resistance device [illegible] two [illegible]

5.9 Determine [illegible] of an idealized transistor [illegible] to Fig. [illegible] and 5.2.

5.10 [illegible] Fig. Ex5.[illegible]

[illegible]

[illegible]

6

Large signal effects

The exact physical and mathematical description of nonlinear effects related to applied large signals is a complicated procedure, at least if taking into account complex device parameters. Special computer programs have to be used, which are not discussed here. Only some effects of practical interest will be treated briefly. Whether or not a given input signal leads to considerable nonlinear effects depends on the device and its current–voltage characteristics.

As mentioned in Chapters 1 and 2, the assumption of linear behaviour of the circuit is only true if sufficiently small signals are applied, e.g. across a p–n or M–S junction sinusoidal signals with amplitudes

$$\hat{V} < V_{\mathrm{T}} = \frac{kT}{q} \tag{6.1}$$

(temperature voltage $V_{\mathrm{T}} \approx 26\,\mathrm{mV}$ at room temperature) or in the case of a given transfer function

$$i = \sum_{\nu=0}^{\infty} a_{\nu} v^{\nu} \tag{6.2}$$

for sufficiently small parameters a_{ν} $(\nu > 1)$; otherwise nonlinear effects occur, the signals become distorted and saturation effects have to be taken into account (e.g. Heiter, 1973). Contrary to linear distortion, which is of importance for small signal pulse transmission because of amplitude and phase deformations, nonlinear effects lead to amplitude-dependent bias conditions and generation of harmonics. This is applicable to rectification or mixing and frequency multiplication. In these cases the variation of the amplitude-dependent transfer behaviour should be as large as possible. On the other hand, if linear transfer is to be afforded, the bias point has to be chosen in the linear part of the transfer characteristic, and negative feedback has to be applied to smooth the resulting characteristic further.

6.1 SWITCHING

The large signal switching behaviour of diodes is treated at the end of section 5.2.1. It is easily extendable to three-terminal active devices (e.g. Le Can *et al.*, 1962).

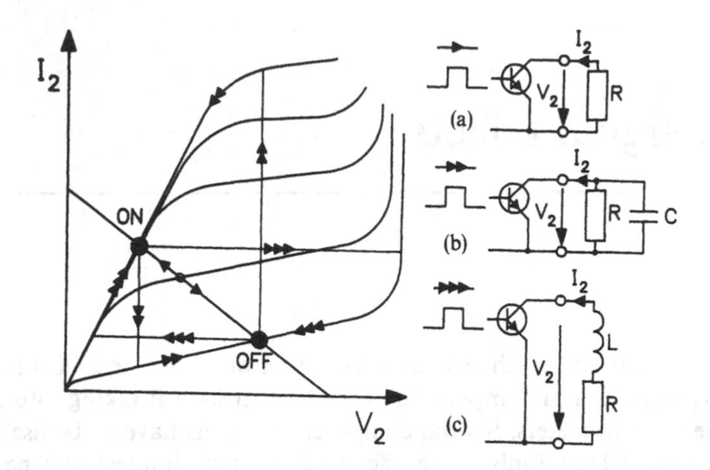

Fig. 6.1 Idealized output switching characteristics. (a) Resistive load; (b) capacitive load; (c) inductive load.

Here the large switching behaviour from an 'on' state to an 'off' state and *vice versa* of an active device will be treated under the assumption that the device itself can be switched relatively fast by the input signal, and that the outer output circuit determines the time behaviour. This common situation results in dangerous overcurrents and overvoltages, which can destroy the device. They are dependent on the characteristic current–voltage dependence at the output of the device itself and on the amplitude and steepness of the applied input signal. In Fig. 6.1 the idealized switching path at the output of the active device is indicated for ohmic load (a) and for complex loads (b), (c). The area enclosed is relevant for the power consumption per switching cycle. It follows that because of safety reasons the driving input signal should not be too fast and large.

In the 'on' condition, if a bipolar transistor is driven into saturation, the accumulation of charge in the base layer resulting from the forward biased collector p–n junction additionally leads to a delayed turn-off (e.g. Beneking, 1989; Le Can *et al.*, 1962; Sze, 1990). The relevant storage time t_s corresponds to p–n diodes (section 5.2.1, and Exercises 6.1 and 6.3).

A reduced storage effect is observable for bipolar transistors exhibiting a wide-gap collector (n–p–N type, p–n–P type) because of the high inner barrier. If a Schottky collector is used instead of a p–n junction this storage time can be totally avoided because the M–S diode does not inject charges from the metal in the base if forward biased. Figure 6.2 gives an example for both cases (Beneking and Su, 1982; Beneking *et al.*, 1980).

If in contrast to the limiting outer circuit the active device is relevant for the switching behaviour, the transfer function output (current) to input (voltage) has to be discussed. This functional dependence is of importance for active devices

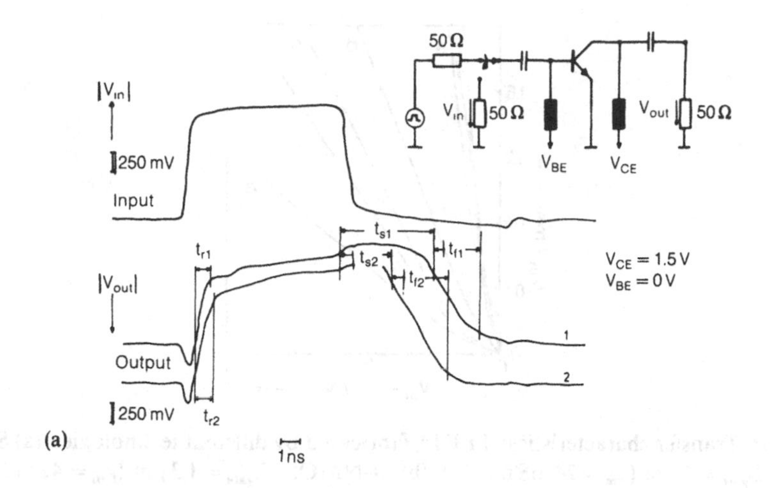

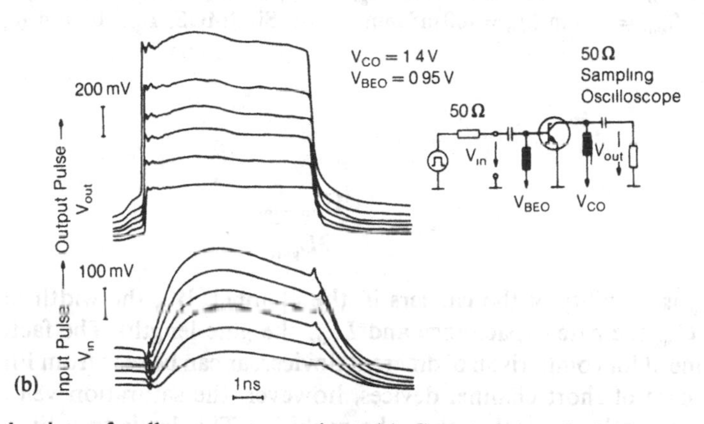

Fig. 6.2 Reduction of collector storage time. (a) Comparison of N–p–n and N–p–N widegap heterojunction GaAs bipolar transistors (After Beneking and Su, 1982.) (b) Schottky-collector GaAs heterojunction bipolar transistor (N–p–M). (After Beneking *et al.*, 1980.)

used in digital circuits for memory functions, etc., and allows the comparison of different technologies and types of devices. The steeper this transfer is the faster the turn-on (or turn-off) results, assuming a ramp input signal.

For a rough estimate a quadratic dependence

$$I_{out} = K_1(V_{in} - V_0)^2 \tag{6.3}$$

can be assumed, where e.g. in case of a MISFET $V_0 = V_{th}$ is the threshold voltage

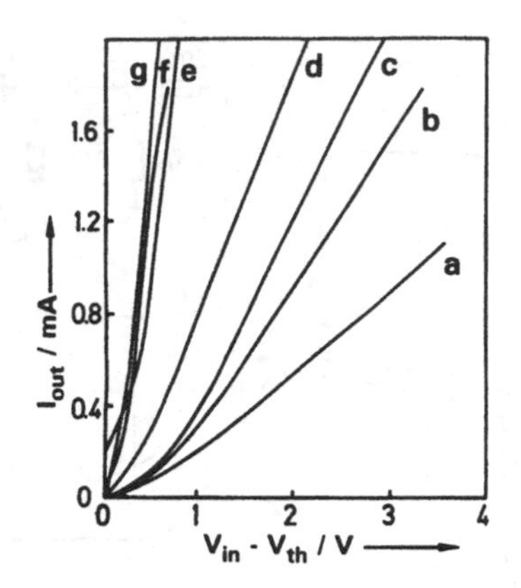

Fig. 6.3 Transfer characteristics of FETs, fabricated by different technologies. (a) Si-MESFET, $L_{gate} = 1\,\mu m$ ($g_m = 24\,mS\,mm^{-1}$); (b) Si-NMOS, $L_{gate} = 1.3\,\mu m$ ($g_m = 42\,mS\,mm^{-1}$); (c) Si-NMOS, $L_{gate} = 0.7\,\mu m$ ($g_m = 60\,mS\,mm^{-1}$); (d) Si-NMOS, $L_{gate} = 0.5\,\mu m$ ($g_m = 130\,mS\,mm^{-1}$); (e) GaInAs-MISFET, L_{gate} $1.5\,\mu m$ ($g_m = 300\,mS\,mm^{-1}$); (f) GaAs-MESFET, $L_{gate} = 0.5\,\mu m$ ($g_m = 400\,mS\,mm^{-1}$); (g) Si-NMOS, L_{gate} $0.1\,\mu m$ ($g_m = 520\,mS\,mm^{-1}$).

and

$$K_1 = \mu_{ch} \frac{W_{ch} C_{ins}}{2L_{gate}} \tag{6.4}$$

where μ_{ch} is mobility of the carriers in the channel, W_{ch} the width of the FET channel, C_{ins} the gate capacitance and L_{gate} the gate length). The factor K_1 is a factor of merit for comparison of different devices, as can be seen from Fig. 6.3.

In the case of short channel devices, however, the saturation velocity v_{sat} of the carriers is relevant rather than the mobility. This leads to a linear transfer function

$$I_{out} = K_2(V_{in} - V_0) \tag{6.5}$$

with

$$K_2 = v_{sat} W_{ch} C_{ins} \tag{6.6}$$

In Fig. 6.3 the dependencies of curves (f) and (g) show clearly the more linear behaviour for low input voltages.

6.2 POWER AMPLIFICATION

Power amplification of RF signals is an extension of small signal amplification to large signals, where impedance matching becomes difficult. The reasons are

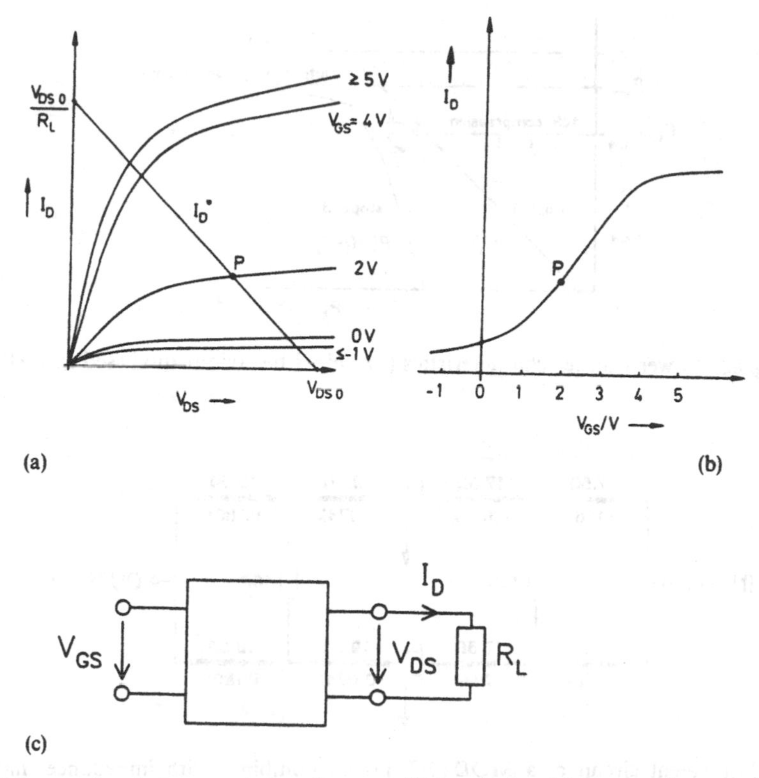

Fig. 6.4 Large signal characteristics of an amplifying device. (a) Output characteristic with ohmic load line I_D^*; (b) transfer characteristic; (c) black box verification.

the amplitude-dependent complex parameters of the amplifying device which lead to different optimum conditions for different signal amplitudes. Furthermore, because of the nonlinear characteristics and transfer of the input signal, saturation occurs if large signals are applied.

Assuming large signal behaviour of a FET as shown in Fig. 6.4, the nonlinear current response to the applied voltage becomes evident. From the load line (I_D^*) the clink factor can be determined, either experimentally or mathematically. By negative feedback a reduction is achievable (section 4.1.1). In any case, a saturation will be present, and the output power becomes a nonlinear function of the input power.

A characteristic figure of merit is the 1 **dB compression point** which is defined as the output power P_{out} in decibels (dBm, related to 1 mW) where a reduction in the power gain of 1 dB is found (Fig. 6.5). This value depends of course on the circuit configuration used (bias, etc.).

Because of the relatively low inner impedances of semiconductor power devices, matching sections have to be implemented to reduce the impedance of the outer connections (stripline, coplanar line, waveguide, finline (Frey and Bhasin, 1985;

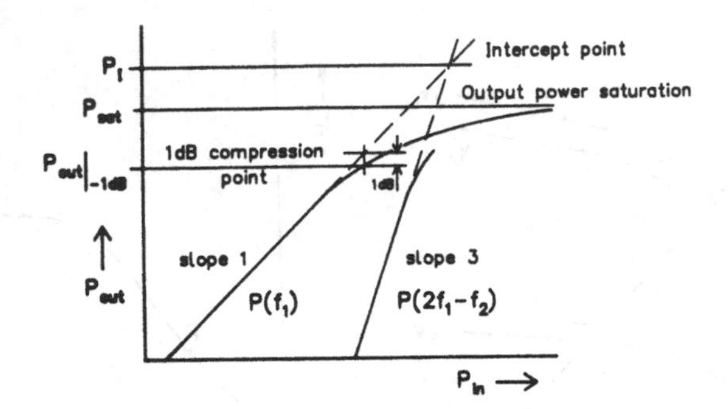

Fig. 6.5 Power transfer characteristics (principal behaviour discussed in text).

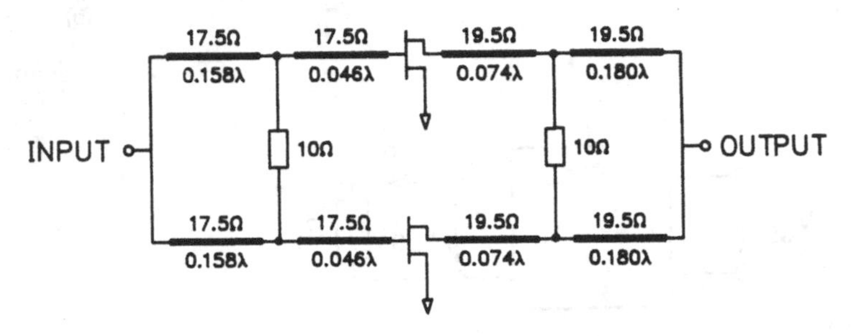

Fig. 6.6 Equivalent circuit of a MODFET power combiner with impedance matching (After Arai *et al.*, 1991.)

Pucel, 1985)). It should be mentioned here that in many cases for hybrid or monolithic MICs, coplanar lines are preferable because they allow short grounding pads of the active devices and other lumped elements directly at the surface of the substrate used.

To couple active devices together to achieve higher power, special power splitters and combiners are used. Figure 6.6 gives an example, where two MODFET chips are combined in connection with Wilkinson-type transformers to realize a 55 GHz hybrid integrated amplifier (Arai *et al.*, 1991). The resulting frequency response is shown in Fig. 6.7, whereas output power and efficiency are shown in Fig. 6.8.

The resulting output power saturation corresponding to Fig. 6.5 can clearly be seen. The efficiency is known as the **power added efficiency**

$$\eta_a = \frac{(P_{out} - P_{in})}{P_{bias}} \tag{6.7}$$

This quantity is better suited for the characterization of power amplifiers in

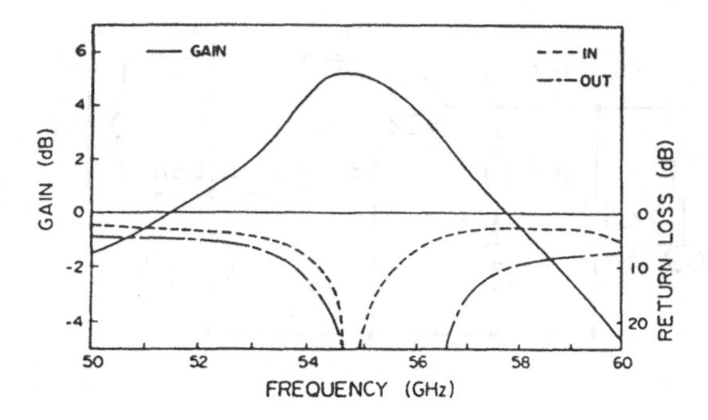

Fig. 6.7 Measured frequency response of gain and return loss (input power −5 dBm). (After Arai *et al.*, 1991.)

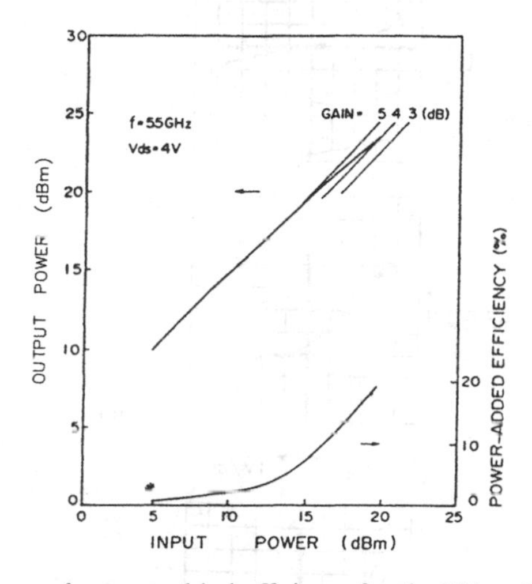

Fig. 6.8 Output power and power-added efficiency for the U-band amplifier with pseudomorphic GaInAs power MODFETS. (After Arai *et al.*, 1991.)

comparison with the conventional power efficiency

$$\eta = \frac{P_{\text{out}}}{P_{\text{bias}}} \tag{6.8}$$

because the RF input power is also included.

In the case of a narrowband RF power amplifier with a resonant circuit at the output (Fig. 6.9(a)), another effect has to be mentioned. In the case of a large voltage swing at the output a direct output–input connection through the device becomes possible, which links the connected circuit temporarily at the input side

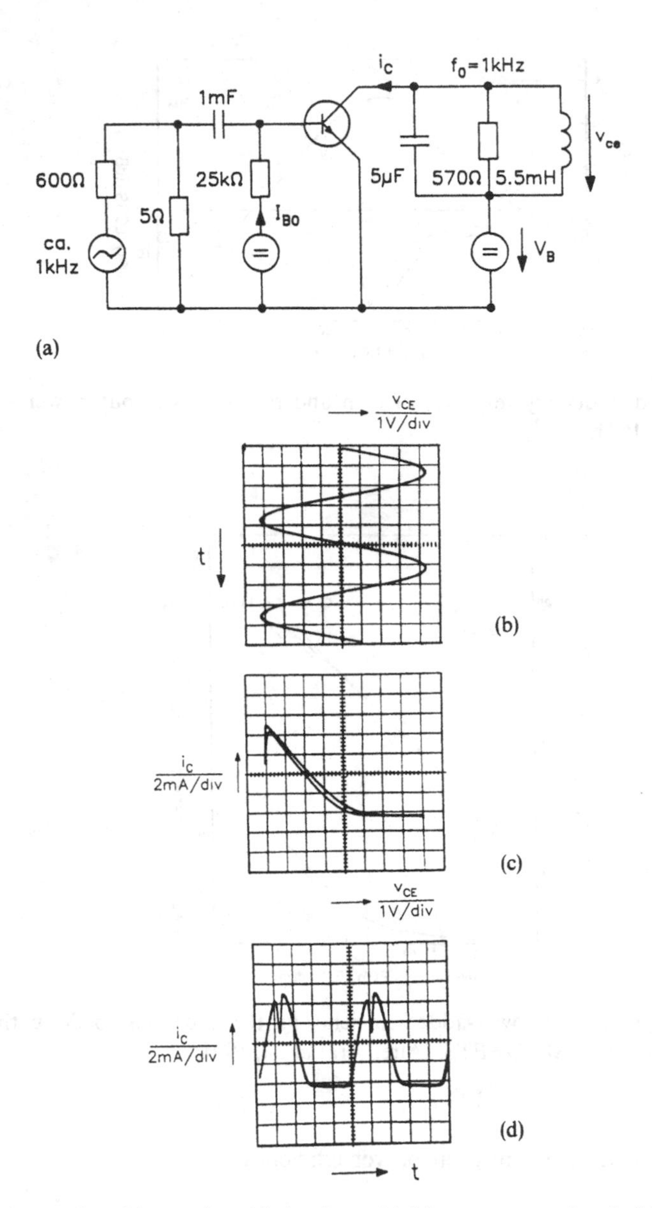

Fig. 6.9 Narrowband large signal behaviour. (a) Measurement circuit (bipolar transistor GFT32/30) (b) sinusoidial output voltage; (c) dynamic characteristic (exceeding the DC load line) (d) collector current. (After Thoma, 1969.)

of the device to the output. The effect is pronounced for bipolar transistors but relevant for all devices where an inner connection between output and input exists.

If, as is common, a relatively large input signal is applied, current saturation occurs at the output with a more or less rectangular waveform similar to large signal switching. The connected LC circuit selects the first harmonics only, leading to a sinusoidal output voltage (Fig. 6.9(b)). Its swing is larger than the voltage corresponding to the load line between the 'off' and 'on' condition because the first Fourier coefficient of a rectangular waveform is $4/\pi$. Therefore the effective load line is extended to higher voltages along the limiting output characteristic of the device and to lower voltages along the differential low-ohmic part of the characteristics before saturation (Fig. 6.9(c)). In this region not only does the output current drop (Fig. 6.9(d)), but also the collector diode becomes positively biased, resulting in a low-ohmic inner connection to the input side. As a result the circuit behaviour at the output is linked to the input, and strong differences follow for low generator impedances at the input (voltage source) and high generator impedances (current source). Whereas, e.g. for fast switching of a bipolar transistor a voltage source with low inner impedance is preferable, in case of an RF power amplifier a sufficiently high impedance connected to the input is needed.

The selection of the bias condition is also of importance regarding the achievable efficiency or maximum output power. If a resonant circuit is used relatively large nonsinusoidal input signals can be used, allowing effective amplification (predistorted or rectangular wave function at the input). Symmetrical stages with relatively low bias currents are preferable in this respect.

6.3 NONLINEAR DISTORTION

The distortion of signals by nonlinear response of active (and passive) devices is difficult to describe. Many influencing parameters exist, which are dependent on the signal amplitude itself, and complex signal variations in amplitude and phase occur. Special computer programs are available, but the determination of the relevant parameters is not easy. Therefore semirigorous methods have been developed, and in many cases only Taylor series are discussed (e.g. Minasian, 1980).

Two groups are of interest, time domain programs (e.g. SPICE, MICROCAP) and frequency domain programs (e.g. LIBRA, MICROWAVE HARMONICA). The latter include the linear part of a circuit in the frequency domain, whereas the nonlinear part is treated in the time domain (harmonic balance method (e.g. Chang *et al.*, 1990). These analytical CAD tools are not covered here. Instead the current components at the resulting frequencies $f_\nu = \nu\omega/2\pi$ are given corresponding to a Taylor series, if a voltage signal $v(t) = \hat{V}\cos\omega t$ is applied. Including third-order effects, it is

$$i = a_1 v + a_2 v^2 + a_3 v^3 \tag{6.9}$$

It follows, after ordering and neglecting higher order terms ($|a_3|\hat{V}^2 \ll |a_1|$), that

$$i = \frac{a_2}{2}\hat{V}^2 + a_1\hat{V}\cos\omega t + \frac{a_2}{2}\hat{V}\cos 2\omega t + \frac{a_3}{4}\hat{V}^3\cos 3\omega t \qquad (6.10)$$

As can be seen, the coefficient a_2 is relevant for rectification and frequency doubling, etc. It is also relevant for mixing, if two RF signals with frequencies f_1 and f_2 are applied.

However, nonlinear distortion is not only an important issue for power amplifiers with large input signals. In the case of front ends the cross-modulation is a critical effect. This is the superimposing of the modulation of a strong signal, also received at the first stage of an amplifier, at a frequency near that of the weak input signal to be received (e.g. Perlow, 1976).

If the distorting large input signal is given by

$$v_S = \hat{V}_S(1 + m_S \sin\Omega t)\sin\omega_S t \qquad (6.1)$$

and the signal to be received by

$$v_1 = \hat{V}_1 \sin\omega_1 t \qquad (6.12)$$

then the modulation m transferred on to the primarily unmodulated carrier becomes

$$m = \frac{3a_3\hat{V}_S^2 m_S}{a_1 + \frac{3}{4}a_3\{\hat{V}_S^2(2 + m_S^2) + \hat{V}_1^2\}} \qquad (6.13)$$

which under usual conditions ($\hat{V}_1 \ll \hat{V}_S$, $|a_3|\hat{V}_S^2 < |a_1|$) leads to

$$m = m_S\frac{3a_3}{a_1}\hat{V}_S^2 \qquad (6.14)$$

The coefficients a_1, a_3 belong to the Taylor series of the corresponding nonlinear device characteristic as given above,

$$i = a_1 v + a_2 v^2 + a_3 v^3 + \cdots \qquad (6.15)$$

Therefore front ends should exhibit linear or quadratic transfer functions to avoid this cross-modulation effect. Because, in contrast to bipolar transistors, FETs show a nearly quadratic transfer behaviour, they are preferable devices in this respect.

To determine the cross-modulation behaviour of a device a two-frequency method is used. As shown in Fig. 6.5, besides the power transfer of the original signal ($f = f_1$), the power is also measured at a frequency $f_m = 2f_1 - f_2$, where f_2 is chosen in such a manner that f_m and f_2 are also located in the amplified channel, $f_m, f_2 \approx f_1$. An advanced measuring setup for the measurement of intermodulation distortion at GHz frequencies is given in Rauscher and Tucker (1977).

It follows from the analysis, assuming the same effective load resistor R_1, that the output powers are, respectively,

$$P(f_1) = \frac{a_1^2 \hat{V}_1^2}{2R_1} \tag{6.16}$$

and

$$P(f_m) = \frac{9a_3^2 \hat{V}_1^4 \hat{V}_S^2}{32R_1} \tag{6.17}$$

Therefore the ratio m/m_S becomes

$$\frac{m}{m_S} = \frac{3a_3}{a_1} \hat{V}_S^2 = \left(\frac{P(f_m)}{P(f_1)}\right)^{1/2} \tag{6.18}$$

if $\hat{V}_S = \hat{V}_1$ is chosen.

As can be concluded from the power equations, the slope of $P(f_1)$ in a log–log plot of $P_{in}(f_1)$ exhibits a unit slope, whereas that of $P(f_m)$ shows a slope of three as long as the input exhibits the same input resistance independent of the frequency. Therefore the extrapolation of both curves leads to the characteristic intercept point P_I (Fig. 6.5), which is a figure of merit regarding the third-order distortion (the higher the value of P_I the lower the distortion). From the result shown in Fig. 6.10 it becomes evident that the buried channel FET is advantageous in this respect (Beneking *et al.*, 1982).

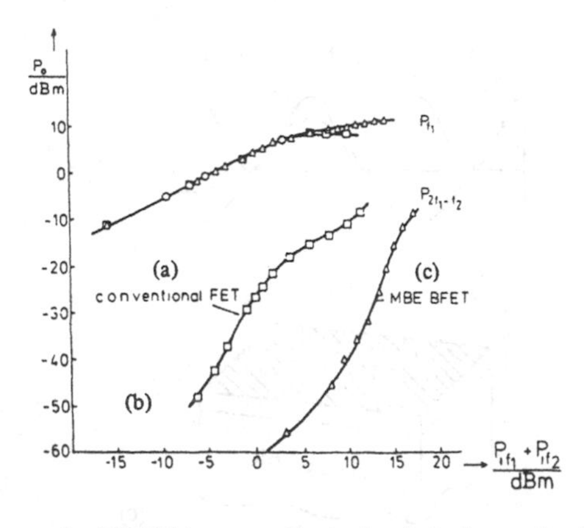

Fig. 6.10 Comparison of a BFET (grown GaAs buried channel) with a conventional FET (a) Fundamental output power P_{f1} and ((b), (c)) the third-order intermodulation product $P_{2f_1-f_2}$ as a function of the input power $P_{if_1} + P_{if_2} = 2P_{if_1}$. (After Beneking *et al.*, 1982.)

6.4 DYNAMIC TEMPERATURE EFFECTS

Generally it is assumed that the applied RF signal does not change the device temperature periodically; only a steady state temperature increase is expected, depending on the mean value of the absorbed power

$$\overline{P} = \langle p(t) \rangle = \frac{\overline{[v(t)]^2}}{R} \tag{6.19}$$

In this equation

$$v(t) = \hat{V} \cos \omega t \tag{6.20}$$

is the voltage applied to the resistance R (Fig. 6.11(a)). This leads to

$$p(t) = \frac{\hat{V}^2}{R}\left(\tfrac{1}{2} + \tfrac{1}{2}\sin 2\omega t\right) \tag{6.21}$$

In general the alternating part is neglected, and

$$\overline{P} = \frac{\hat{V}^2}{2R} \tag{6.22}$$

is the commonly used expression.

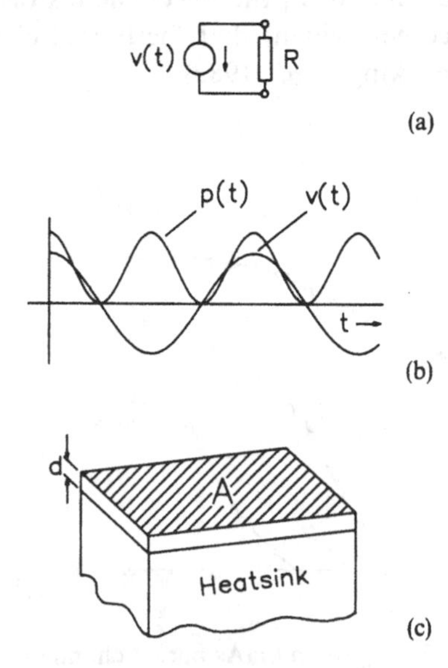

Fig. 6.11 Dynamic temperature behaviour. (a) Circuit; (b) alternating voltage and power; (c) geometry.

As can be seen the absorbed power $p(t)$ varies with $f_p = 2f = \omega/\pi$ (Fig. 6.11(b)). This temporal change can influence the device parameters. However, this remarkable effect will only occur if the frequency f of the applied RF signal is low enough to allow a corresponding temperature change in the affected volume.

An exact solution of the problem can be found by solving the time-dependent diffusion equation (e.g. Kittel and Krömer, 1980). However, an estimate can be made by comparing the time constants involved. If the thermal time constant τ_{th} of the volume influenced exceeds the period length of the power oscillation,

$$\tau_{th} > \frac{1}{2\pi f_p} \tag{6.23}$$

then no disturbance will occur. If, however,

$$\tau_{th} \lesssim \frac{1}{2\pi f_p} \tag{6.24}$$

then an effect is expected.

As an example a planar configuration will be considered (Fig. 6.11(c)). This consists of a conducting channel, area A and thickness d, on a heatsink with low thermal resistance. Then the thermal time constant can be expressed by

$$\tau_{th} = R_{th} C_{th} = \frac{c_v \rho_v}{\sigma_{th}} d^2 \tag{6.25}$$

where

$$R_{th} = \frac{d}{\sigma_{th} A} \tag{6.26}$$

and

$$C_{th} = c_v \rho_v V_p \tag{6.27}$$

The active volume, where the power absorption takes place, is

$$V_p = Ad \tag{6.28}$$

where σ_{th} is the thermal conductivity, c_v the specific heat and ρ_v the specific mass. The critical signal frequency f_c is therefore

$$f_c = \frac{1}{4\pi\tau_{th}} = \frac{1}{4\pi} \frac{\sigma_{th}}{c_v \rho_v} \frac{1}{d^2} \tag{6.29}$$

For $f < f_c$, thermal relaxation effects will occur if the signal strength is suffi-

Table 6.1 Thermal data at room temperature for Si and GaAs

	Si	GaAs
σ_{th}(W °C^{-1} cm)	1.5	0.46
c_v(J °C^{-1} g)	0.7	0.35
ρ_V(g cm^{-3})	2.33	5.32

ciently high to develop a remarkable temperature change in the affected volume. With the data given in Table 6.1 it follows that for Si,

$$\frac{f_c}{\text{GHz}} \approx 0.7 \frac{1}{\left(\dfrac{d}{\mu\text{m}}\right)^2} \tag{6.30}$$

and for GaAs

$$\frac{f_c}{\text{GHz}} \approx 0.2 \frac{1}{\left(\dfrac{d}{\mu\text{m}}\right)^2} \tag{6.31}$$

Therefore conducting channels of submicrometre thickness can be affected at frequencies in the GHz range. Because σ_{th} becomes smaller at lower temperatures, e.g. 1/5 of the RT value at 77 K, this frequency limit is even lower at low temperatures.

EXERCISES

6.1 Derive simplified analytical expressions for the turn off behaviour of a p–n diode after forward biasing, corresponding to Fig. Ex6.1. The diode is assumed to be a short-base p^+–n diode (base width $W < L_D$, the minority carrier diffusion length of the holes in the n base).

Derive the expressions for the storage time t_s and the (assumed) exponential decay, which depends on the inner time constant of the device and the current ratio I_R/I_F (I_F is the forward current before switching, I_R the reverse current after turn-off at $t = t_S$; the current immediately after turn off is $I_{RO} \approx I_R$).

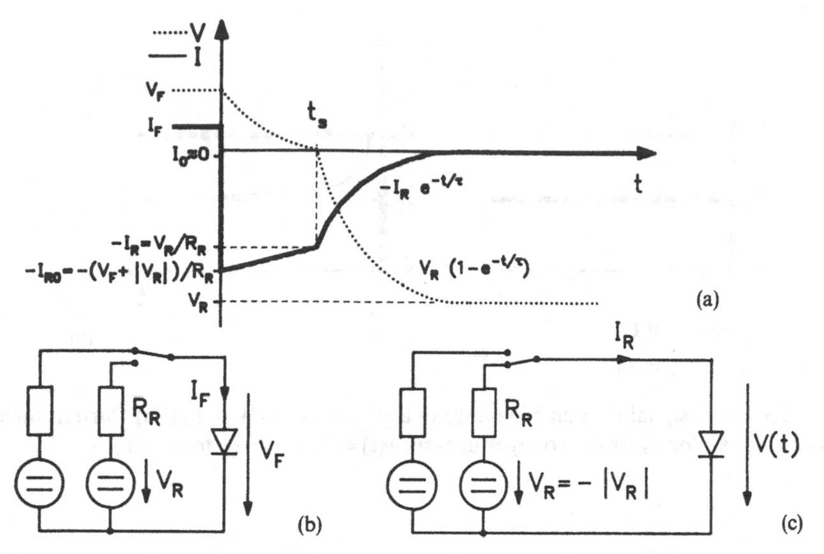

Fig. Ex6.1 Large signal switching of a p–n diode. (a) Current and voltage after application of the reverse voltage; (b) forward-biased diode; (c) reverse-biased diode.

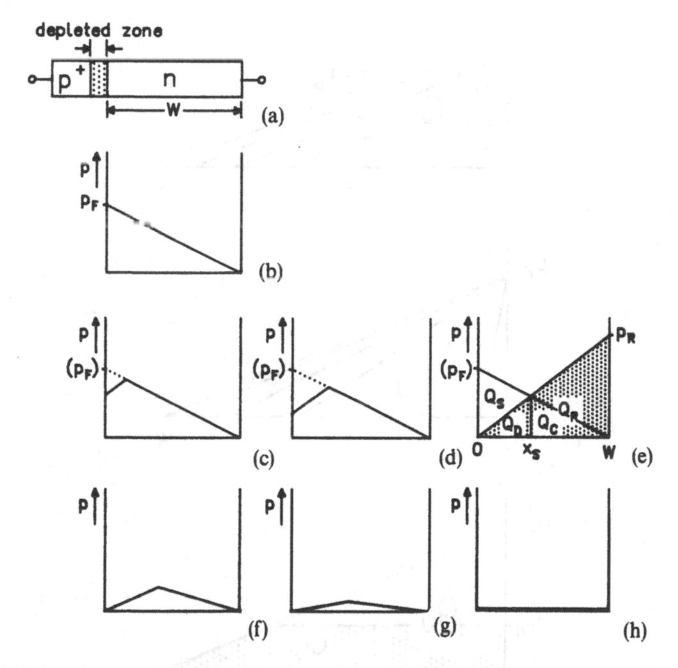

Fig. Ex6.2 Minority carrier distributions in the base (n part) of a p^+–n diode after application of a reverse voltage (from forward direction; the real net charge is zero because of the compensating majority carriers). (a) Diode configuration (short base, $W \ll L_D$); (b) steady state distribution for forward biasing $I_F > 0$; ((c), (d)) intermediate density distributions; (e) situation at $t = t_s$; ((f), (g)) decay phase; (h) resulting steady state $I_R \approx 0$.

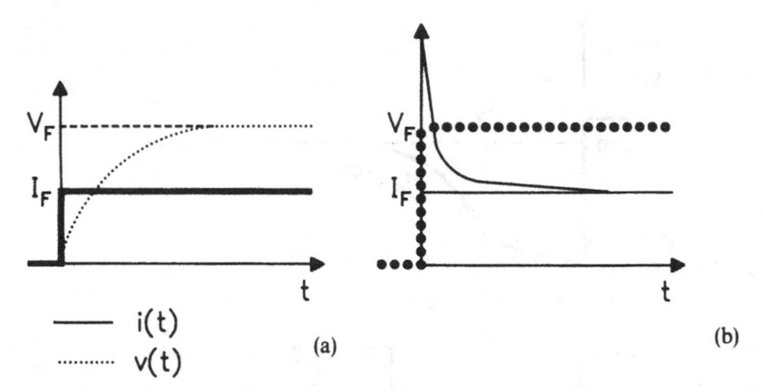

Fig. Ex6.3 Large signal turn on behaviour of a p^+–n diode ($W \ll L_D$). (a) Current turn-on, $i(t) = I_F = \text{const.}$ for $t \geqslant 0$; (b) voltage turn-on, $v(t) = V_F = \text{const.}$ for $t \geqslant 0$.

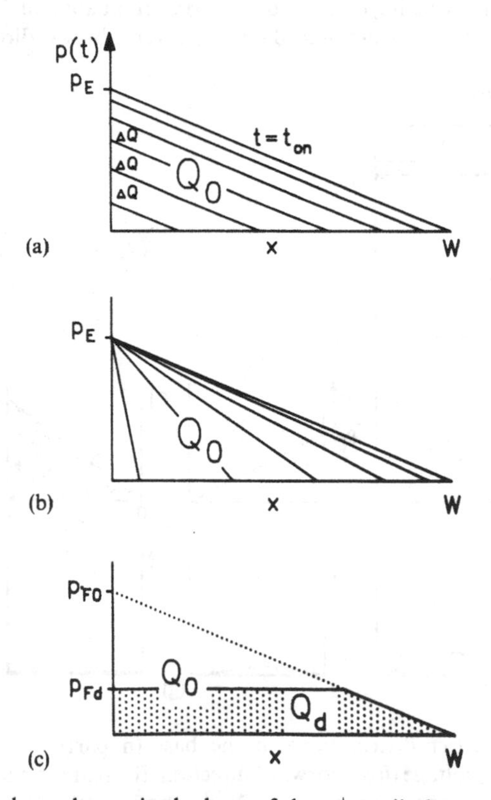

Fig. Ex6.4 Density dependences in the base of the p^+–n diode at turn-on (schematic). (a) Current turn-on; (b) voltage turn-on; (c) steady-state density distribution with built-in drift field.

Draw first in a simplified version the corresponding minority carrier distribution in the base for $i = I_F$ with $p = p_F$ at $x = 0$ and $p = 0$ at $x = W$. Indicate the change of this distribution after application of the reverse voltage V_R (Fig. Ex6.1(c)) and draw the time-dependent hole density for $0 < t \leqslant t_S$ until the value $p(x = 0, t = t_S) = 0$ is reached. Indicate in the diagram the fictional density $p(x = W) = p_R$ for this moment, which is proportional to the extracted reverse current I_R. Derive the expression for t_S dependent on the ratio I_R/I_F.

The second phase, $t > t_S$, is the decay phase, where a dependence of

$$i(t - t_S) = I_R e^{-(t - t_S)/\tau}$$

can be assumed. Determine the characteristic time constant τ.

6.2 Draw simplified diagrams of the time-dependent minority carrier densities $p(x, t)$ in the n base of a p^+–n diode (short base, $W < L_D$), in the case of (Fig. Ex6.3) (a) current turn-on and (b) voltage turn-on. Determine the corresponding turn-on times by taking into account the time-dependent stored carrier densities.

Which final carrier distribution would occur, if a positive drift field is built-in (e.g. by exponential decay of the n-doping in the base)?

6.3 Derive the equation for the determination of the storage time t_S of a bipolar n–p–n transistor if turned off after saturation. Express the saturation factor I_{Ef}/I_{Efsat} by the applied input voltage V_{BE} and the input voltage V_{BEsat} for reaching saturation (Fig. Ex6.5). Indicate the (fictional) gradients of both injection densities n_E, n_C, belonging to the forward emitter and collector currents I_{Ef} and I_{Cf}.

6.4 Discuss the circuit shown in Fig. Ex6.6. It consists of an unconventional driving circuit for a laser diode, to avoid a long tail after excitation. Explain the function of the pulse-shaping circuit. Give the length l of the transmission line depending on the required pulse length Δt.

6.5 Derive the relationship for the cross-modulation effect, if two signals

$$V_1 = v_1 \cos \omega_1 t \text{ and } V_2 = v_2 \cos \omega_2 t \quad \text{with} \quad v_2 = v_{20}(1 + m_2 \cos \Omega t)$$

are applied ($v_{20} \gg v_1$), and the transfer function is given as (section 6.3)

$$i = a_1(V_1 + V_2) + a_2(V_1 + V_2)^2 + a_3(V_1 + V_2)^3$$

Explain the two-frequency method for the determination of the factors

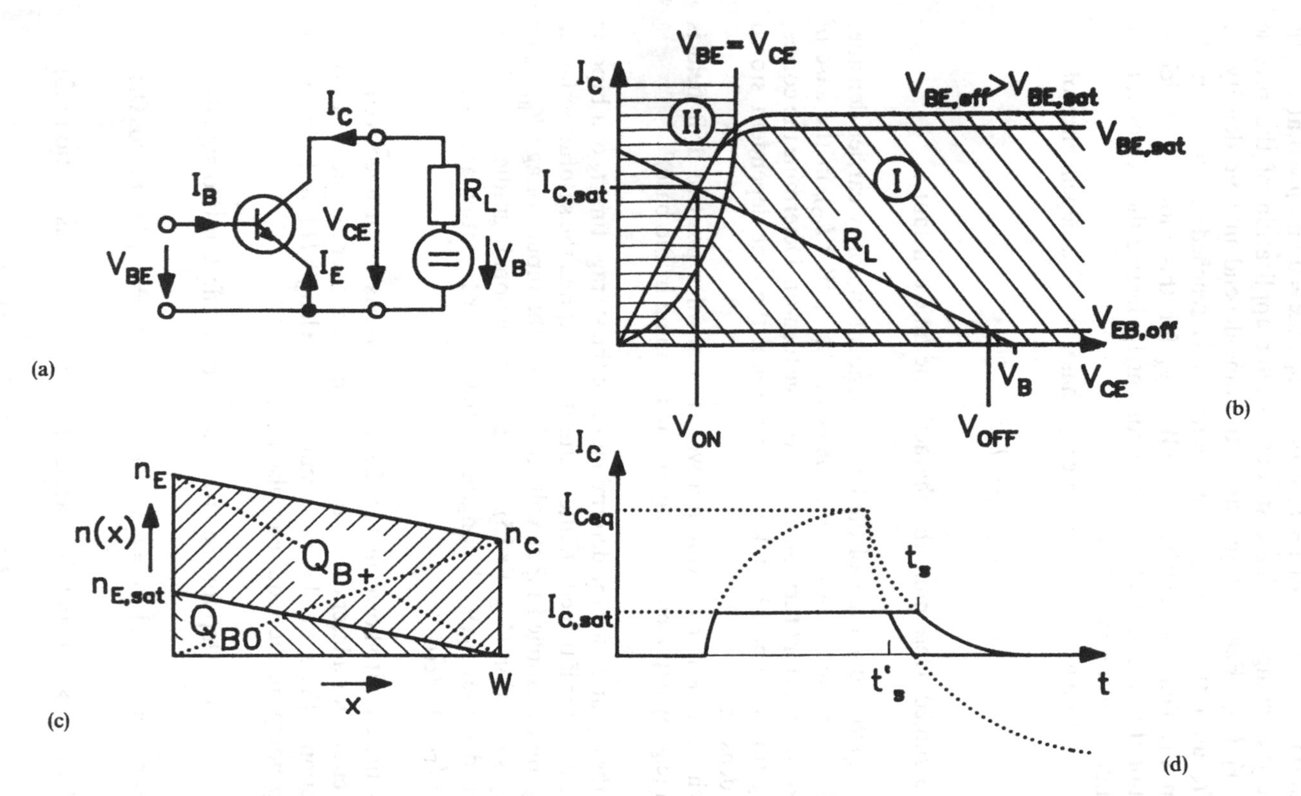

Fig. Ex6.5 Saturation of bipolar transistors (common emitter) and related time delay at turn-off. (a) Circuit configuration; (b) I_C–V_{CE} diagram (region I normal active region with forward-biased EB junction and reverse-biased CB junction; region II saturation region with forward biasing of both junctions, $I_{Ef} \sim n_E$, $I_{Cf} \sim n_C$); (c) idealized minority carrier distribution (Q_{B0} base charge at the saturation point, limit of region I; Q_B+ extra charge additive to Q_{B0} if saturation takes place, region II); (d) $I_C(t)$ dependence.

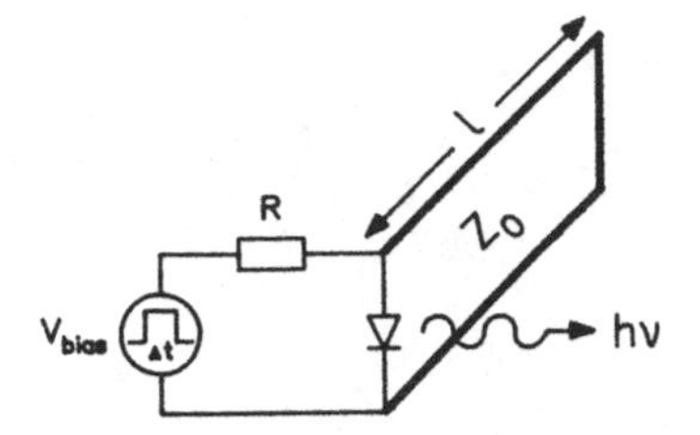

Fig. Ex6.6 Circuit with transmission line to reduce the tail of the emitted light pulse of a luminescent diode (laser diode).

a_v of the transfer function. (An advanced measuring set-up for the determination of the intermodulation product is described in Rauscher and Tucker (1977).)

Appendix A
Fourpole and twoport parameters

The most important fourpole parameters are the *y*- and *h*-parameters. At GHz frequencies where voltages and currents become difficult to measure the wave concept is applied, leading to the *S*- and *T*-parameters. Figure A.1 summarizes these parameters in the matrix configuration and indicates the definition of positive directions for the currents, voltages and waves. Figure A.2 demonstrates the use of the fourpole parameters for describing coupled fourpoles. As indicated each set of fourpole parameters is applicable for special fourpole combinations (Table 4.1, Fig. Ex4.4). To characterize one single fourpole any set of parameters can be used, as long as stability exists. In Table A.1 the transformations of the

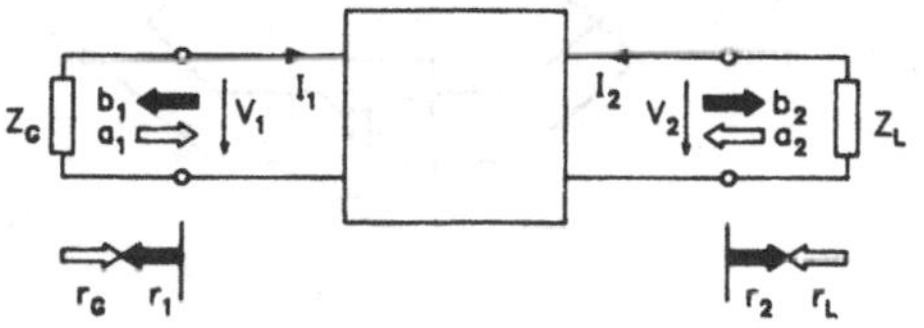

$$a_i = \frac{1}{2}\left(\frac{V_i}{\sqrt{Z_{0i}}} + I_i\sqrt{Z_{0i}}\right) \qquad b_i = \frac{1}{2}\left(\frac{V_i}{\sqrt{Z_{0i}}} - I_i\sqrt{Z_{0i}}\right)$$

$$\frac{1}{Y_{0i}} = Z_{0i} \text{ transmission line impedance}$$

$$\begin{pmatrix} I_1 \\ I_2 \end{pmatrix} = (Y)\begin{pmatrix} V_1 \\ V_2 \end{pmatrix}$$ Admittance matrix parameters y_{kl}

$$\begin{pmatrix} V_1 \\ I_2 \end{pmatrix} = (H)\begin{pmatrix} I_1 \\ V_2 \end{pmatrix}$$ Hybrid matrix parameters h_{kl}

$$\begin{pmatrix} b_1 \\ b_2 \end{pmatrix} = (S)\begin{pmatrix} a_1 \\ a_2 \end{pmatrix}$$ Scattering matrix parameters S_{kl}

$$\begin{pmatrix} b_1 \\ a_1 \end{pmatrix} = (T)\begin{pmatrix} a_2 \\ b_2 \end{pmatrix}$$ Transmission matrix parameters T.

Fig. A.1 Matrix equations of important fourpole and twoport parameters.

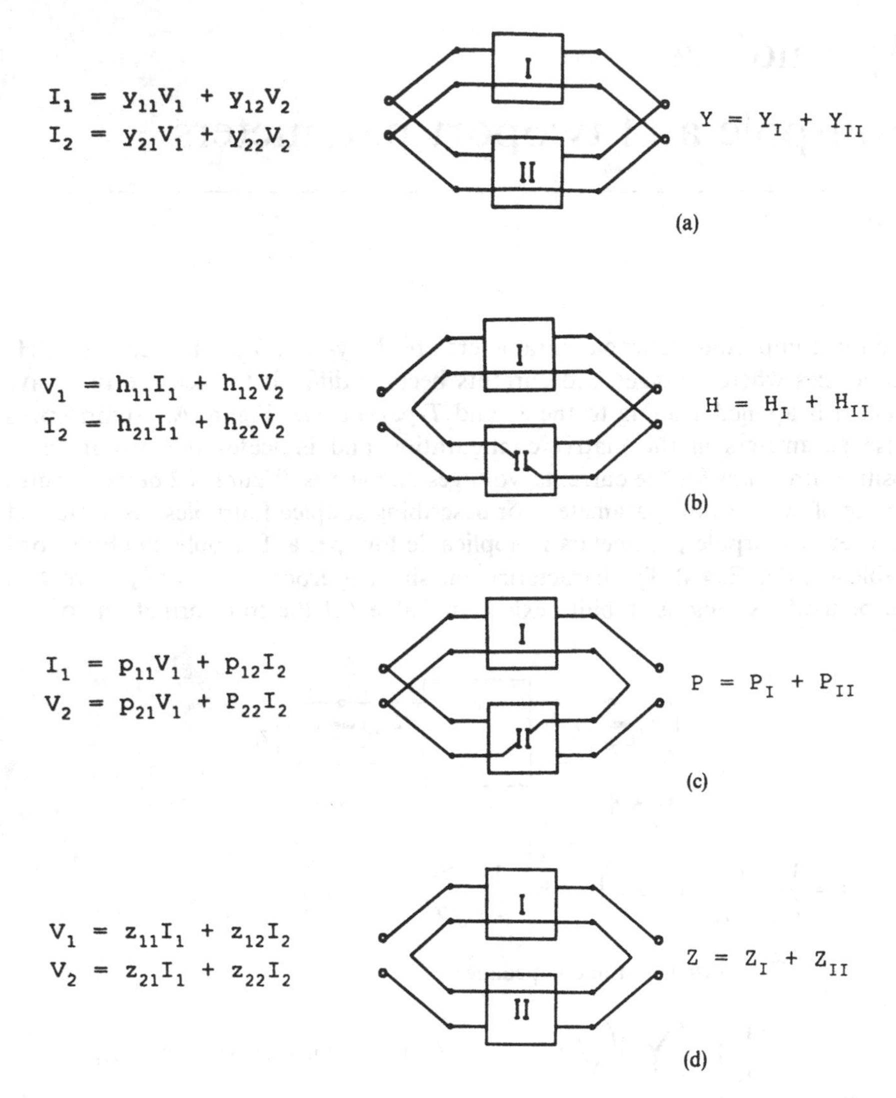

Fig. A.2 Fourpole equations and coupled fourpoles. (a) Parallel–parallel coupling; (b) Series–parallel coupling; (c) Parallel–series coupling; (d) Series–series coupling.

fourpole parameters are listed. The conversion of the y-parameters into the wave parameters and *vice versa* is given in Table A.2.

As an example, the short-circuit current gain h_{21} is expressed by using S-parameters. Because

$$h_{21} = \frac{y_{21}}{y_{11}} \tag{A.1}$$

Table A.1 Conversion of fourpole parameters

	$y_{11}y_{12}y_{21}y_{22}$	$z_{11}z_{12}z_{21}z_{22}$
y_{11}		$\frac{z_{22}}{\lvert Z\rvert}=\frac{1}{z_{11}(1-\delta_z)}$
y_{12}	$I_1 = y_{11}V_1 + y_{12}V_2$	$-\frac{z_{12}}{\lvert Z\rvert}=\frac{1}{z_{21}(1-(1/\delta_z))}$
y_{21}	$I_2 = y_{21}V_1 + y_{22}V_2$	$-\frac{z_{21}}{\lvert Z\rvert}=\frac{1}{z_{12}(1-(1/\delta_z))}$
y_{22}		$\frac{z_{11}}{\lvert Z\rvert}=\frac{1}{z_{22}(1-\delta_z)}$
z_{11}	$\frac{y_{22}}{\lvert Y\rvert}=\frac{1}{y_{11}(1-\delta_y)}$	
z_{12}	$-\frac{y_{12}}{\lvert Y\rvert}=\frac{1}{y_{21}(1-(1/\delta_y))}$	$V_1 = z_{11}I_1 + z_{12}I_2$
z_{21}	$-\frac{y_{21}}{\lvert Y\rvert}=\frac{1}{y_{12}(1-(1/\delta_y))}$	$V_2 = z_{21}I_1 + z_{22}I_2$
z_{22}	$\frac{y_{11}}{\lvert Y\rvert}=\frac{1}{y_{22}(1-\delta_y)}$	
p_{11}	$\frac{\lvert Y\rvert}{y_{22}}=y_{11}(1-\delta_y)$	$\frac{1}{z_{11}}$
p_{12}	$\frac{y_{12}}{y_{22}}$	$-\frac{z_{12}}{z_{11}}$
p_{21}	$-\frac{y_{21}}{y_{22}}$	$\frac{z_{21}}{z_{11}}$
p_{22}	$\frac{1}{y_{22}}$	$\frac{\lvert Z\rvert}{z_{11}}=z_{22}(1-\delta_z)$
h_{11}	$\frac{1}{y_{11}}$	$\frac{\lvert Z\rvert}{z_{22}}=z_{11}(1-\delta_z)$
h_{12}	$-\frac{y_{12}}{y_{11}}$	$\frac{z_{12}}{z_{22}}$
h_{21}	$\frac{y_{21}}{y_{11}}$	$-\frac{z_{21}}{z_{22}}$
h_{22}	$\frac{\lvert Y\rvert}{y_{11}}=y_{22}(1-\delta_y)$	$\frac{1}{z_{22}}$

$|Y| = y_{11}y_{22} - y_{12}y_{21}$; $|Z| = z_{11}z_{22} - z_{12}z_{21}$

$\delta_y = \frac{y_{12}y_{21}}{y_{11}y_{22}}$; $\delta_z = \frac{z_{12}z_{21}}{z_{11}z_{22}}$

Table A.1 (Cont.)

	$p_{11}p_{12}p_{21}p_{22}$	$h_{11}h_{12}h_{21}h_{22}$
y_{11}	$\frac{\|P\|}{p_{22}} = p_{11}(1-\delta_p)$	$\frac{1}{h_{11}}$
y_{12}	$\frac{p_{12}}{p_{22}}$	$-\frac{h_{12}}{h_{11}}$
y_{21}	$-\frac{p_{21}}{p_{22}}$	$\frac{h_{21}}{h_{11}}$
y_{22}	$\frac{1}{p_{22}}$	$\frac{\|H\|}{h_{11}} = h_{22}(1-\delta_h)$
z_{11}	$\frac{1}{p_{11}}$	$\frac{\|H\|}{h_{22}} = h_{11}(1-\delta_h)$
z_{12}	$-\frac{p_{12}}{p_{11}}$	$\frac{h_{12}}{h_{22}}$
z_{21}	$\frac{p_{21}}{p_{11}}$	$-\frac{h_{21}}{h_{22}}$
z_{22}	$\frac{\|P\|}{p_{11}} = p_{22}(1-\delta_p)$	$\frac{1}{h_{22}}$
p_{11}		$\frac{h_{22}}{\|H\|} = \frac{1}{h_{11}(1-\delta_h)}$
p_{12}	$I_1 = p_{11}V_1 + p_{12}I_2$	$-\frac{h_{12}}{\|H\|} = \frac{1}{h_{21}(1-(1/\delta_h))}$
p_{21}	$V_2 = p_{21}V_1 + p_{22}I_2$	$-\frac{h_{21}}{\|H\|} = \frac{1}{h_{12}(1-(1/\delta_h))}$
p_{22}		$\frac{h_{11}}{\|H\|} = \frac{1}{h_{22}(1-\delta_h)}$
h_{11}	$\frac{p_{22}}{\|P\|} = \frac{1}{p_{11}(1-\delta_p)}$	
h_{12}	$-\frac{p_{12}}{\|P\|} = \frac{1}{p_{21}(1-(1/\delta_p))}$	$V_1 = h_{11}I_1 + h_{12}V_2$
h_{21}	$-\frac{p_{21}}{\|P\|} = \frac{1}{p_{12}(1-(1/\delta_p))}$	$I_2 = h_{21}I_1 + h_{22}V_2$
h_{22}	$\frac{p_{11}}{\|P\|} = \frac{1}{p_{22}(1-\delta_p)}$	

$|P| = p_{11}p_{22} - p_{12}p_{21}$; $|H| = h_{11}h_{22} - h_{12}h_{21}$

$\delta_p = \frac{p_{12}p_{21}}{p_{11}p_{22}}$; $\delta_h = \frac{h_{12}h_{21}}{h_{11}h_{22}}$

Table A.2 Conversion of wave parameters

$$\begin{bmatrix} y_{11} & y_{12} \\ y_{21} & y_{22} \end{bmatrix} = \frac{1}{1+S_{11}+S_{22}+\Delta S}\begin{bmatrix} (1-S_{11}+S_{22}-\Delta S)Y_{01} & -2S_{12}(Y_{01}Y_{02})^{1/2} \\ -2S_{21}(Y_{01}Y_{02})^{1/2} & (1+S_{11}-S_{22}-\Delta S)Y_{02} \end{bmatrix}$$

$$\Delta S = S_{11}S_{22} - S_{12}S_{21}$$

$$\begin{bmatrix} S_{11} & S_{12} \\ S_{21} & S_{22} \end{bmatrix} = \frac{1}{1+\dfrac{y_{11}}{Y_{01}}+\dfrac{y_{22}}{Y_{02}}+\dfrac{\Delta y}{Y_{01}Y_{02}}}\begin{bmatrix} 1-\dfrac{y_{11}}{Y_{01}}+\dfrac{y_{22}}{Y_{02}}-\dfrac{\Delta y}{Y_{01}Y_{02}} & -\dfrac{2y_{12}}{(Y_{01}Y_{02})^{1/2}} \\ -\dfrac{2y_{21}}{(Y_{01}Y_{02})^{1/2}} & 1+\dfrac{y_{11}}{Y_{01}}-\dfrac{y_{22}}{Y_{02}}-\dfrac{\Delta y}{Y_{02}Y_{02}} \end{bmatrix}$$

$$\Delta y = Y_{11}Y_{22} - Y_{12}Y_{21}$$

$$\begin{bmatrix} T_{11} & T_{12} \\ T_{21} & T_{22} \end{bmatrix} = \begin{bmatrix} S_{12}-\dfrac{S_{11}S_{22}}{S_{21}} & \dfrac{S_{11}}{S_{21}} \\ -\dfrac{S_{22}}{S_{21}} & \dfrac{1}{S_{21}} \end{bmatrix}$$

$$\begin{bmatrix} S_{11} & S_{12} \\ S_{21} & S_{22} \end{bmatrix} = \begin{bmatrix} \dfrac{T_{12}}{T_{22}} & T_{11}-\dfrac{T_{12}T_{21}}{T_{22}} \\ \dfrac{1}{T_{22}} & -\dfrac{T_{21}}{T_{22}} \end{bmatrix}$$

Y_{01} and Y_{02} are the characteristic admittances of the transmission lines at the input and output, respectively.

it follows that

$$h_{21} = \frac{-2S_{21}(Y_{01}Y_{02})^{1/2}}{(1-S_{11}+S_{22}-\Delta S)Y_{01}} \tag{A.2}$$

If the characteristic impedances are the same at input and output, $Z_{01} = Z_{02}$ and therefore $Y_{01} = Y_{02}$,

$$h_{21} = \frac{2S_{21}}{S_{11}S_{22}-S_{12}S_{21}+S_{11}-S_{22}-1} \tag{A.3}$$

Appendix B
The wave source

The wave source is the wave equivalent of a current or voltage source in the conventional fourpole notation. To introduce this concept we will consider the configuration shown in Fig. B.1(a). If the source is connected to a transmission line, characteristic impedance Z_0, then for $Z_G \neq Z_0$ reflection takes place. If a wave a_G enters the source part,

$$b_G = r_G a_G \tag{B.1}$$

If the source is an active wave source, then it transmits into the line a source wave b_Q, which is superimposed on the reflected wave $r_G a_G$, leading to

$$b_G = r_G a_G + b_Q \tag{B.2}$$

If connected to the input of the following twoport, which exhibits an input port reflection

$$r_1 = \frac{b_1}{a_1} \tag{B.3}$$

then (Fig. B.1(b))

$$b_G = a_1, a_G = b_1 \tag{B.4}$$

From these equations it follows that

$$a_1 = b_Q \frac{1}{1 - r_1 r_G} \tag{B.5}$$

and

$$b_1 = b_Q \frac{1}{1 - r_1 r_G} r_1 \tag{B.6}$$

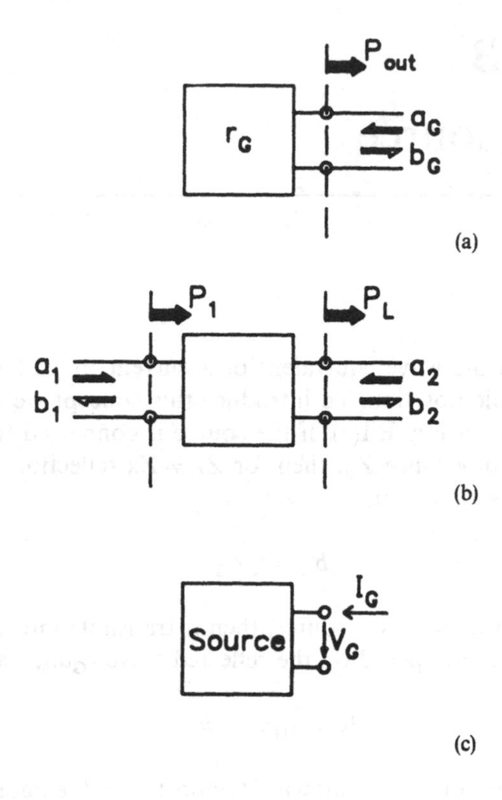

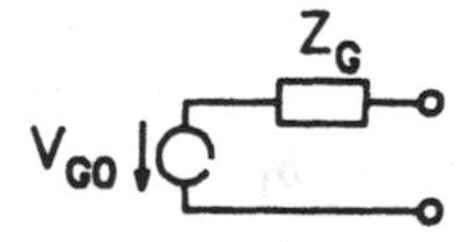

Fig. B.1 The wave source concept. (a) Wave source connection; (b) twoport, connected to the source; (c) definition of positive directions of I_G, V_G; (d) equivalent voltage generator.

The input power of the twoport is

$$P_1 = \frac{|a_1|^2}{2} - \frac{|b_1|^2}{2} \tag{B.7}$$

Substituting the expression for a_1 and b_1,

$$P_1 = \frac{|b_Q|^2}{2} \frac{1 - |r_1|^2}{|1 - r_1 r_G|^2} \tag{B.8}$$

If $r_1 = 0$, then

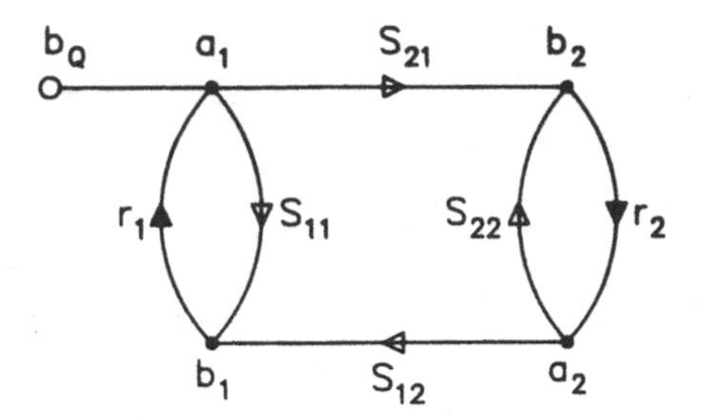

Fig. B.2 Signal flow graph of a twoport connected to a wave source.

$$P_1|_{r_1=0} = \frac{|b_Q|^2}{2} \tag{B.9}$$

If however, $r_1 \neq 0$, then the optimum value of P_1 is achieved for

$$r_1 = r_G^* \tag{B.10}$$

This leads to

$$P_{1\max} = P_1|_{r_1=0} = \frac{|b_Q|^2}{2} \frac{1}{1-|r_1|^2} \tag{B.11}$$

which is the maximum absorbable power at the input of the twoport for a generator with $r_G \neq 0$. Therefore this quantity is the available power of the generator,

$$P_{1\max} = P_{AVG} \tag{B.12}$$

The wave source followed by an active twoport (Fig. B.1(b) acts as a wave-amplifying system. The evaluation of the gain

$$G = \frac{P_L}{P_{AVG}} \tag{B.13}$$

can be performed by using the signal flow graph in Fig. B.2, and is not covered here.

Appendix C Unilateralization

As shown by Rollet (1962) the circuit of Fig. 2.3 allows lossless neutralization for any fourpole. Corresponding to Appendix A the circuit consists of the series connection of the active fourpole with the reactance $X_1 = 1/B_1$, connected in parallel with the fourpole consisting of $X_2 = 1/B_2$. The whole circuit can be described by the matrix equation

$$Y_\Sigma = (Z_S)^{-1} + Y_P \tag{C.1}$$

if

$$Z_S = \begin{bmatrix} z_{11} & z_{12} \\ z_{21} & z_{22} + jX_1 \end{bmatrix} \tag{C.2}$$

describes the active fourpole $Z = (z_{kl})$ together with X_1 at the output, and

$$Y_P = \begin{bmatrix} jB_2 & -jB_2 \\ -jB_2 & jB_2 \end{bmatrix} \tag{C.3}$$

represents X_2.

For neutralization the feedback parameter $y_{\Sigma 12}$ has to be zero, which consists of the sum of

$$y_{S12} = \frac{-z_{S12}}{\Delta z_S} \quad \text{and} \quad -jB_2 \tag{C.4}$$

$$\Delta z_S = z_{S11} z_{s22} - z_{S12} z_{S21} \tag{C.5}$$

Therefore

$$\frac{-z_{21}}{z_{11} z_{22} + jz_{11} X_1 - z_{12} z_{21}} - jB_2 = 0 \tag{C.6}$$

is the condition for neutralization.

Appendix D
Instability and oscillation

As mentioned in section 4.1.1 the critical value $\phi_b = 1/G_0$ marks the instability point, or $K = 1$ as shown in section 2.2. To discuss the time-dependent growth of the corresponding signals the complete differential equation of the underlying electronic circuit has to be treated.

Three regions can be distinguished allowing separate solutions. The first region is the linear region where the signals can be handled as small signals, and the linearized equivalent circuit of all devices involved can be applied. The second region is the nonlinear region, where a mathematical treatment becomes difficult. The third region is the saturation region, where a steady state situation is reached, and where again simple conclusions can be drawn. The first region is the domain of the complex frequencies $p = \sigma + j\omega$ and shall be treated here.

It is not the aim of this section to introduce the Fourier and Laplace transformation or to give an introduction of the circuit theory in general. The reader is referred to the pioneering work of Nyquist (1932) regarding the mathematical treatment of stability and instability, and also to Kuo (1966) and de Pian (1962) in respect of the circuit theory involved. Only the practical application of complex frequencies to the study of the stability situation from a circuit point of view will be presented here.

Any electrical circuit consists of passive components and active sources leading to the common equivalent small signal circuit. If a given point of this circuit is disconnected, then an external voltage source can be applied leading to the principal circuit configuration shown in Fig. D.1(a). Correspondingly a dual version can be verified if, in parallel to two nodes of the circuit, an external current source is connected (Fig. D.1(c)). As mentioned in section 1.4, those sources exhibit no inner impedance nor admittance, respectively, and therefore no further passive components are added to the original linear circuit.

From Fig. D.1(a) it follows for the superimposed current that

$$I_1 = \frac{V_{10}}{Z_1} \tag{D.1}$$

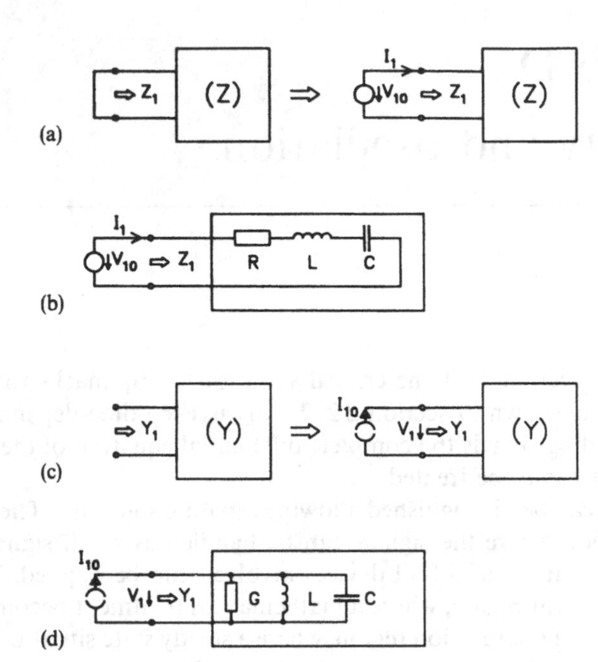

Fig. D.1 Circuit configurations. (a) Impedance behaviour; (b) Characteristic impedance system; (c) Admittance behaviour; (d) Characteristic admittance system.

and for the superimposed voltage (Fig. D.1(c))

$$V_1 = \frac{I_{10}}{Y_1} \tag{D.2}$$

where Z_1, Y_1 are the impedance and admittance of the network, respectively, at the connection port.

The supplied complex power is

$$P = \frac{I_1^*}{2^{1/2}} \frac{V_{10}}{2^{1/2}} = \frac{|I_1|^2}{2} Z_1 \tag{D.3}$$

and

$$P = \frac{V_1^*}{2^{1/2}} \frac{I_{10}}{2^{1/2}} = \frac{|V_1|^2}{2} Y_1 \tag{D.4}$$

respectively. For simplicity a very simple circuit configuration is assumed, a series-connected damped oscillatory circuit (Fig. D.1(b)), and the dual version, a parallel-connected *RLC* circuit (Fig. D.1(d)).

Therefore

$$Z_1 = R + j\omega L + \frac{1}{j\omega C} \tag{D.5}$$

and

$$Y_1 = G + j\omega C + \frac{1}{j\omega L} \tag{D.6}$$

respectively.

D.1 CHARACTERISTIC EQUATION

To introduce the characteristic equations, the corresponding differential equation which is relevant to the system

$$j\omega \rightarrow p = \sigma + j\omega \tag{D.7}$$

has to be set, as is common in the case of the Laplace transformation.

Without supplied power into the circuit, $P = 0$, in the first case a current $I_1 \neq 0$ is only possible for $Z(p) = 0$, and in the second case a voltage V_1 can only be developed for $Y(p) = 0$.

These two equations are the characteristic equations belonging to the differential equation of the system. The roots p_ν, p_μ of $Z(p) = 0$ and $Y(p) = 0$, respectively, correspond to the solutions

$$I_1 = \sum_1^n I_{1\nu} = \sum_1^n I_{1\nu o} e^{p_\nu t} \tag{D.8}$$

and

$$V_1 = \sum_1^m V_{1\mu} = \sum_1^m V_{1\mu_o} e^{p_\mu t} \tag{D.9}$$

of the homogeneous part of the differential equation, whereas any driving signal V_{10}, I_{10} is relevant to the inhomogeneous part.

The proof of stability or instability consists of the inspection of the real parts σ_ν, σ_μ of the roots p_ν and p_μ, respectively. If $\sigma_\nu, \sigma_\mu < 0$ the circuit is stable and reacts, depending on the magnitude of the components, after excitation, with a damped oscillation, or in the case of supercritical damping by an aperiodic exponential decay, if at $t = 0$ a driving pulse (I_{10}, V_{10}) is applied. If however, $\sigma_\nu, \sigma_\mu > 0$ then an increasing magnitude of the current and/or voltage occurs, and the system is unstable.

Therefore stability exists if all p_ν, p_μ have a real part $\sigma < 0$, and instability occurs for $\sigma > 0$. The larger these quantities the faster the growth of the corresponding signal after its excitation. The initial amplitudes depend on the condition of the system at $t = 0$. If a δ-type excitation is assumed (Dirac pulse) then all frequencies are activated with the same amplitude.

The situation will be exemplified by treating the simple oscillatory circuit shown in Fig. D.1(d), where the conductance G represents the output conductance G_o of an active device parallel to a conductance, $G = G_o + G_L$.

The introduction of p for $(j\omega)$ in the admittance, $Y(j\omega) \rightarrow Y(p)$, leads to

$$Y(p) = G + pC + \frac{1}{pL} \tag{D.10}$$

This is the characteristic equation of the underlying differential equation, and the roots p_μ of $Y(p) = 0$ lead to the characteristic complex frequencies of the system.

Because this equation is quadratic in p, two solutions are found,

$$p_1 = -\frac{G}{2C} + \left(\left(\frac{G}{2C}\right)^2 - \frac{1}{LC}\right)^{1/2} \tag{D.11}$$

and

$$p_2 = -\frac{G}{2C} - \left(\left(\frac{G}{2C}\right)^2 - \frac{1}{LC}\right)^{1/2} \tag{D.12}$$

The roots are located in the complex p-plane as indicated in Fig. D.2. The arrows show the trends of p_1 and p_2, if G is changing from large positive values to negative ones.

For

$$\left(\frac{G}{2C}\right)^2 < \frac{1}{LC} \tag{D.13}$$

two conjugated complex solutions are found,

$$p_1 = -\frac{G}{2C} + j\left(\frac{1}{LC} - \left(\frac{G}{2C}\right)^2\right)^{1/2} \tag{D.14}$$

and

$$p_2 = -\frac{G}{2C} - j\left(\frac{1}{LC} - \left(\frac{G}{2C}\right)^2\right)^{1/2} \tag{D.15}$$

If $G > 0$, the system is unconditionally stable, as then in any case σ_1, σ_2 remain negative. Any initial perturbation decays. Therefore all roots located in the left

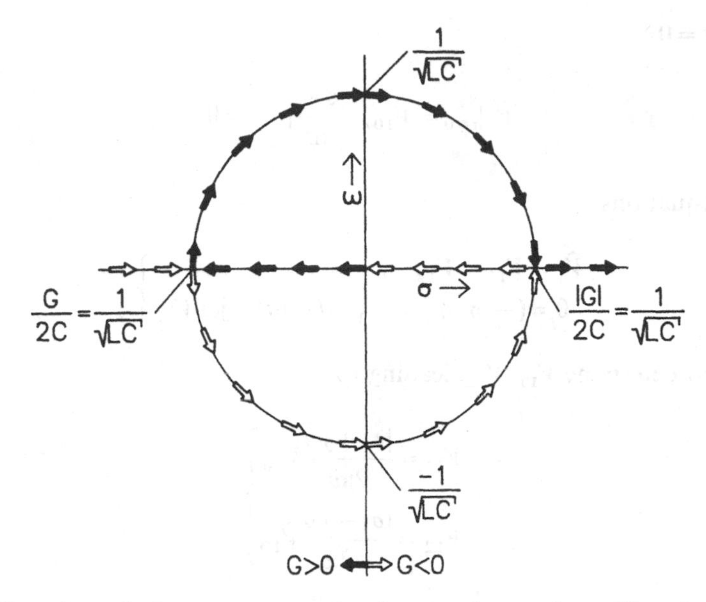

Fig. D.2 Location of the roots p_1, p_2 in the complex p-plane. The charge in p_1 (➡) and p_2 (⇒) is shown for G changing from $+\infty$ to $-\infty$.

hand side of the complex p-plane, as indicated in Fig. D.2, belong to stable circuits.

However, if $G<0$ the σ-values become positive, and instability occurs, in the form either of oscillation or of switching behaviour. For $\sigma=0$ an oscillation with $f=1/(2\pi(LC)^{1/2})$ results with constant amplitude, where again the magnitude depends on the initial condition.

Concerning the roots in the complex p-plane (Fig. D.2), the following situations can occur.

(a) Stability, $G>0$

(i) Oscillatory behaviour Oscillation occurs if

$$\frac{1}{(LC)^{1/2}} > \frac{G}{2C} \tag{D.16}$$

With

$$|\sigma| = \frac{G}{2C}, \quad \omega_0 = \frac{1}{(LC)^{1/2}}, \quad \omega = (\omega_0^2 - \sigma^2)^{1/2}$$

the damped oscillation becomes

$$V_1 = \mathrm{e}^{-|\sigma|t}(V_{11}\,\mathrm{e}^{j\omega t} + V_{12}\,\mathrm{e}^{-j\omega t}) \tag{D.17}$$

If e.g. at $t = 0$

$$V_1|_{t=0} = \hat{V}_{10}, \quad \left.\frac{dV_1}{dt}\right|_{t=0} = 0 \tag{D.18}$$

then the equations

$$\left.\begin{aligned} \hat{V}_1 &= V_{11} + V_{12} \\ 0 &= (-|\sigma| + j\omega)V_{11} + (-|\sigma| - j\omega)V_{12} \end{aligned}\right\} \tag{D.19}$$

allow us to determine V_{11}, V_{12}, leading to

$$\left.\begin{aligned} V_{11} &= \frac{|\sigma| + j\omega}{2j\omega}\hat{V}_{10} \\ V_{12} &= \frac{|\sigma| - j\omega}{-2j\omega}\hat{V}_{10} \end{aligned}\right\} \tag{D.20}$$

Therefore the solution is the damped oscillation

$$v_1(t) = \hat{V}_{10}\,e^{-|\sigma|t}\left(1 + \left(\frac{\sigma}{\omega}\right)^2\right)^{1/2} \sin\left\{\omega t + \arctan\left(\frac{\omega}{|\sigma|}\right)\right\} \tag{D.21}$$

(ii) Critical aperiodic behaviour This occurs if

$$\frac{1}{(LC)^{1/2}} = \frac{G}{2C} \tag{D.22}$$

Then $p_1 = p_2 - |\sigma|$ and

$$V_1 = (V_{11} + V_{12})e^{-|\sigma|t} \tag{D.23}$$

If again

$$V_1|_{t=0} = \hat{V}_{10} \tag{D.24}$$

then an exponential decay

$$v_1(t) = \hat{V}_{10}\,e^{-|\sigma|t} \tag{D.25}$$

occurs, without oscillations.

(iii) Aperiodic behaviour This occurs if

$$\frac{1}{(LC)^{1/2}} < \frac{G}{2C} \tag{D.26}$$

With

$$\Omega = (\sigma^2 - \omega_0^2)^{1/2} \tag{D.27}$$

it follows that

$$v_1(t) = e^{-|\sigma|t}(V_{11}e^{\Omega t} + V_{12}e^{-\Omega t}) \tag{D.28}$$

where again from the initial condition the quantities V_{11}, V_{12} follow.

(b) Undamped oscillation, $G = 0$

In the case of continuous oscillation, $\sigma = 0$,

$$V_1 = V_{11}e^{j\omega_0 t} + V_{12}e^{-j\omega_0 t} \tag{D.29}$$

where, e.g. for

$$V_1|_{t=0} = \hat{V}_{10}, \quad \left.\frac{dV_1}{dt}\right|_{t=0} = g \tag{D.30}$$

it follows that

$$V_{11} = \frac{1}{2}\left(\hat{V}_{10} + \frac{g}{j\omega_0}\right) \tag{D.31}$$

$$V_{12} = \frac{1}{2}\left(\hat{V}_{10} - \frac{g}{j\omega_0}\right) \tag{D.31}$$

Therefore the sinusoidial oscillation

$$v_1(t) = \hat{V}_{10}\cos\omega_0 t + \frac{g}{\omega_0}\sin\omega_0 t \tag{D.33}$$

occurs.

(c) Instability, $G < 0$

In this case initial signals are increasing.

(i) Oscillatory behaviour This occurs if

$$\frac{1}{(LC)^{1/2}} > \frac{|G|}{2C} \tag{D.34}$$

This situation corresponds to that of (a) (i), however, instead of $e^{-|\sigma|t}$ the factor $e^{+|\sigma|t}$ is relevant for the behaviour, and an oscillation with growing amplitudes starts at $t = 0$.

(ii) Critical switching This occurs if

$$\frac{1}{(LC)^{1/2}} = \frac{|G|}{2C} \tag{D.35}$$

In this case an exponential turn-on occurs,

$$v_1(t) = \hat{V}_{10}\,\mathrm{e}^{+|\sigma|t} \tag{D.36}$$

if $\hat{V}_{10}$ again has been the initial amplitude.

(iii) Switching This occurs if

$$\frac{1}{(LC)^{1/2}} < \frac{|G|}{2C} \tag{D.37}$$

The turn-on behaviour is modified in comparison to (c) (ii). However, again an increasing signal without oscillation occurs.

D.2 SATURATION AMPLITUDE

As mentioned earlier the real behaviour of a circuit is not linear for all signal amplitudes. Therefore any turn-on with $G < 0$ is reduced while the growth is in process, leading to a final situation with $G = 0$.

To evaluate the maximum amplitude of the output signal the Moeller oscillation characteristic has to be discussed. It consists of two dependencies in the same graph (Fig. D.3). First the large signal transfer characteristics are plotted regarding the oscillation frequency f_p. In the case of the Hartley oscillator, as shown in Exercise 4.5, the output amplitude $\hat{I}_2(f_o)$ as a function of the input $\hat{I}_1(f_o)$ can be plotted. However, the voltage transfer function can also be used. This curve

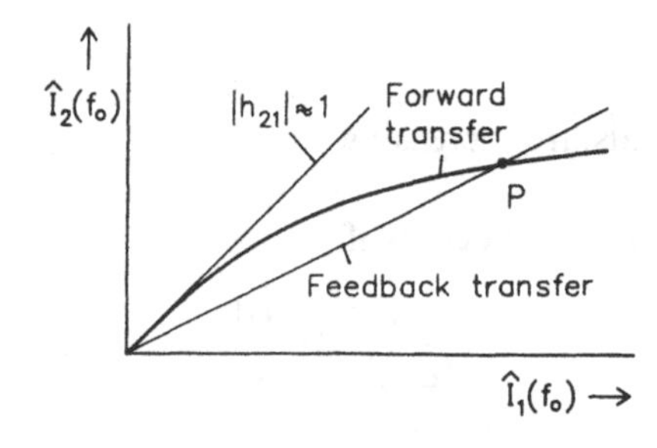

Fig. D.3 Oscillatory characteristics. The crossover point P corresponds to the final amplitudes.

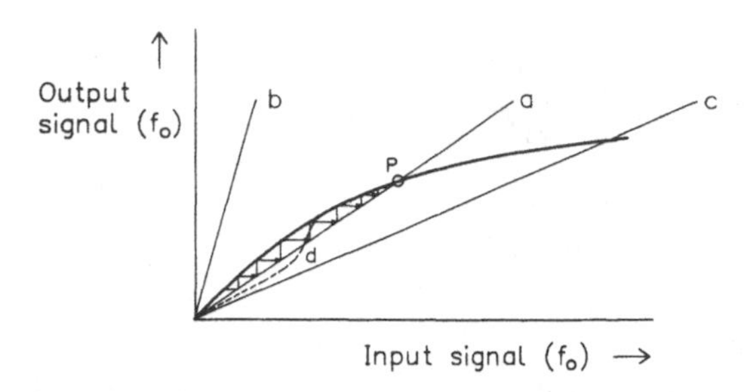

Fig. D.4 General oscillation characteristics of a feedback oscillator. (a) Normal oscillatory behaviour, $\sigma > 0$; (b) Too little feedback, $\sigma < 0$; (c) Extremely high feedback, $\sigma \gg 0$; (d) S-type gain curve, $\sigma < 0$ for small initial amplitudes.

exhibits an initial gradient as given by its small signal value, e.g. $|h_{21}| \approx 1$ (or $|y_{21}|$, respectively). At large amplitudes saturation occurs leading to a maximum value $\hat{I}_{2\max}$ and $\hat{V}_{2\max}$ depending on the active device and its biasing and load conditions. Therefore this first curve shows a tendency to bend as indicated in Fig. D.3, leading to a saturated output signal. The second curve to be drawn is the feedback curve indicating the transfer of the output signal to the entrance port of the active device. Because this transfer is linear and exhibits no saturation, it is represented by a straight line. Its gradient is governed by the feedback coefficient. In the example given in Exercise 4.5 the voltage feedback ratio is chosen as $x_{\text{opt}} = 0.5$, which determines the gradient of the inverse feedback curve $\hat{I}_2/\hat{I}_1$.

Different situations can occur as can be seen from Fig. D.4. The circuit discussed above, Fig. D.3, corresponds to Fig. D.4(a). A stable final amplitude is developed, indicated by the crossover point P. The development of the amplitude follows the stepwise rise as shown in Fig. D.4. The distance between both curves corresponds to σ; the larger it is the faster the growth of the signal. If this distance becomes negative, for weaker feedback or lower gain in the feedback loop, no oscillation will occur (Fig. D.4(b)). On the other hand, for extremely high feedback (Fig. D.4(c)), the strong saturation achieved leads to nonlinear distortions and higher order frequencies. If the gain curve is an S-shape, then a threshold value of the initial amplitude for oscillatory behaviour may exist (Fig. D.4(d)).

Appendix E
Negative impedance circuits and amplifiers

E.1 NEGATIVE IMPEDANCE CONVERTER

In Fig. E.1 a circuit is shown which allows us to convert the connected impedance at the input to the inverse one at the output. The frequency range of these negative impedance converters (NIC) depends on the broadband behaviour of the operational amplifier (OP) and its fourpole parameters (section 5.5.2).

Because of the extremely high gain of the OP the voltages across the resistances R_1 and R_2 have to be (about) the same. With the current flow directions indicated in Fig. E.1,

$$I_1R_1 = -I_2R_2 \tag{E.1}$$

and

$$V_1 = V_2 \tag{E.2}$$

Therefore

$$R_1I_1 = \frac{R_1V_1}{Z_g} = \frac{R_1V_2}{Z_g} = -R_2I_2 \tag{E.3}$$

and

$$Z_2 = \frac{V_2}{I_2} = -Z_g\frac{R_2}{R_1} \tag{E.4}$$

As a result the impedance Z_g at the input becomes transformed to a negative impedance $Z_2 = tZ_g$ at the output, where the transformation factor is

$$t = -\frac{R_2}{R_1} \tag{E.5}$$

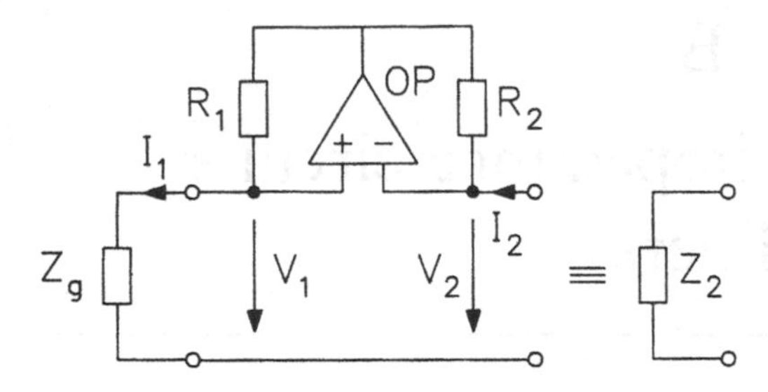

Fig. E.1 Circuit configuration of a negative impedance converter (NIC) using a high gain operational amplifier.

This allows – in a limited frequency range – to compensate for lossy elements and to correct frequency dependencies.

E.2 RF AMPLIFIERS

RF amplification with negative impedances via transmission lines (coaxial lines, striplines, coplanar lines, waveguides, etc.) can be achieved with insertion type and reflection type configurations. In the following, two types are demonstrated.

E.2.1 Tunnel diode amplifier

In Fig. E.2 examples of waveguide-mounted tunnel diodes are shown. In the case of the reflection type, stability is achieved independently of load and generator matching by using a circulator. In the insertion case unidirectional lines on both sides of the amplifier section have to be implemented. In the following a brief description of the (preferable) reflection type is given.

Figure E.3 shows the principal configuration. The circulator itself is omitted, and any losses are neglected; the generator–NR device and NR device–load transmission lines exhibit the same characteristic impedance $Z_0 = 1/G_0$, and also load resistor and generator exhibit the same value, $G_G = G_0, G_L = G_0$.

The incoming power at the port of the NR device (admittance $Y = -Y_-$) is

$$P_{in} = \left(\frac{|V_G|}{2}\right)^2 G_0 \tag{E.6}$$

and the reflected power, which is identical with the power in the load, is

$$P_{out} = |r|^2 P_{in} \tag{E.7}$$

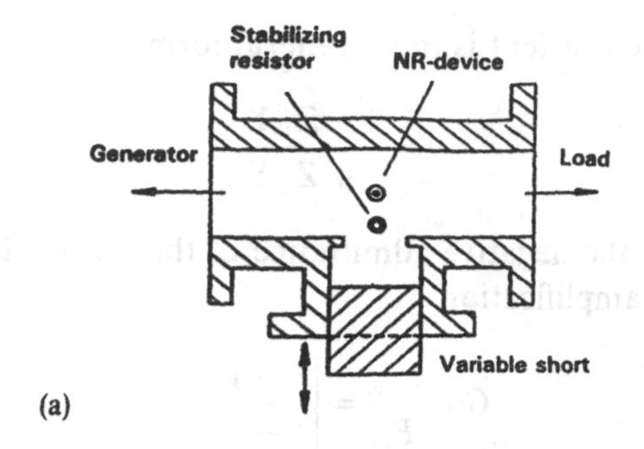

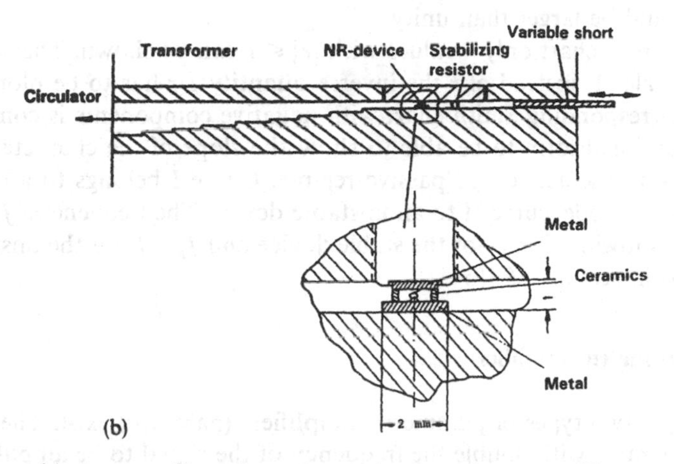

Fig. E.2 Waveguide type configuration of negative impedance amplifiers. (a) Insertion type; (b) Reflection type. (After Henke, 1967.)

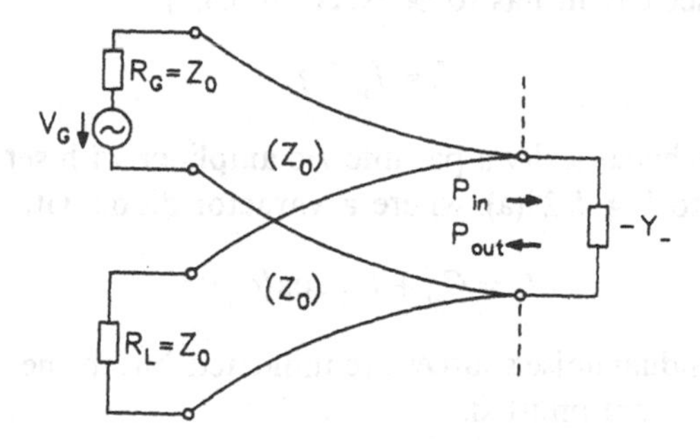

Fig. E.3 Principal configuration of a reflection type NR amplifier.

where the reflection coefficient is in its general form

$$r = \frac{1 - Z_0 Y}{1 + Z_0 Y} \tag{E.8}$$

Taking into account the negative admittance of the NR device, it follows, with $y = -YZ_0$, that the amplification

$$G = \frac{P_{\text{out}}}{P_{\text{in}}} = \left| \frac{1 + y}{1 - y} \right|^2 \tag{E.9}$$

which should be larger than unity.

In the Smith chart only r-values with $|r| \leqslant 1$ can be shown. Therefore, in this case with $|r| > 1$, instead of r the inverse quantity $1/r$ has to be plotted. In Fig. E.4 the corresponding Smith chart with negative components is combined with the conventional form to be able to show the slope of the characteristic of the NR device in the active and passive regime. Curve I belongs to a short-circuit stable tunnel diode, curve II to an unstable device. The frequencies f_a and f_s are indicated, showing $f_a < f_s$ for the stable device and $f_a > f_s$ for the unstable diode, as derived in section 5.5.1(b).

E.2.2 Parametric amplifier

In principle two types of parametric amplifiers (paramps) exist. The first uses a pump generator with double the frequency of the signal to be amplified. Then a strong phase relationship between the pump frequency $f_p = 2f_s$ and the signal frequency f_s must exist. If an idler circuit is used – the most common version – then no such phase relationship is afforded. The resonant frequency f_i of this additional resonance circuit has to be (section 5.5.3)

$$f_i = f_p - f_s \tag{E.10}$$

Figure E.5 shows schematically a parametric amplifier with series resonance LC circuits according to Fig. 5.29(a), where a varactor diode with its capacitance

$$C = C_0 + C_1 \cos(\omega_p t) \tag{E.11}$$

is implemented. Residual noise sources are indicated, but the necessary circulators and the pump circuit are omitted.

Under matched conditions, $R_G = R_L = Z_0$, and resonance the varied capacitance forms a negative resistance

$$-R_N = -\{4\omega_s \omega_i C_1^2 (R_i + R_D)\}^{-1} \tag{E.12}$$

which results in signal amplification.

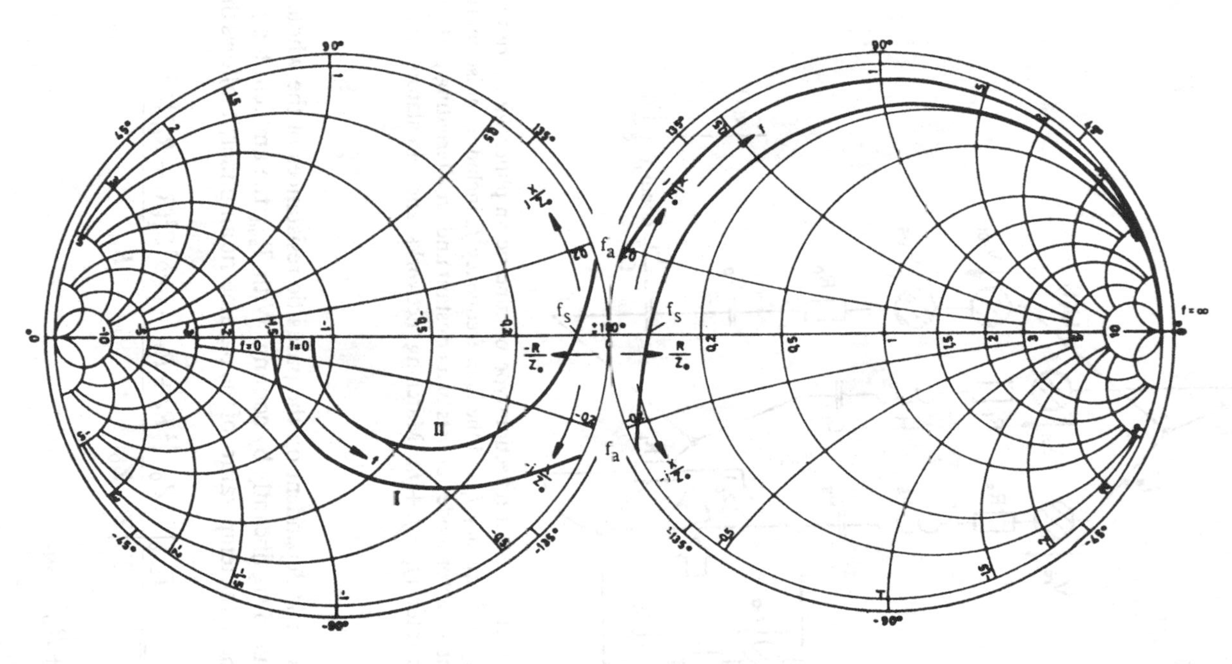

Fig. E.4 Double Smith chart representation of tunnel diode characteristics. I: short-circuit stable device, $f_a < f_s$; II: short-circuit unstable device, $f_a > f_s$. (After Henke, 1967.)

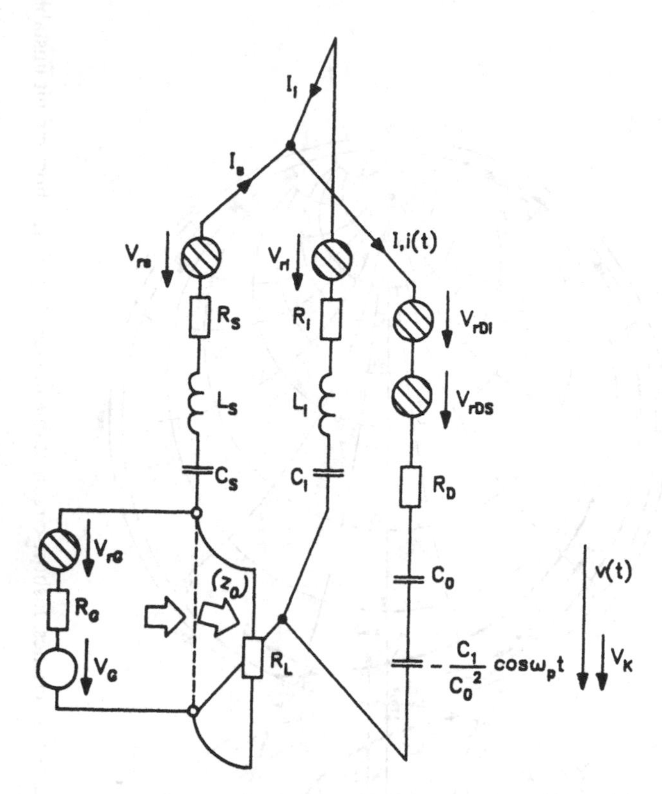

Fig. E.5 Principal circuit for a reflection-type parametric amplifier with series resonant circuits for the signal frequency f_s and the idler frequency f_i, including noise sources. The capacitance of the varactor diode consists of a constant and an alternating part, activated by a pump generator ($f_p = f_s + f_i$). Matching networks and circulators are omitted ($R_G = R_L = Z_0$).

The noise factor is dependent on the parasitic resistances of the whole circuit and can be reduced significantly by cooling. With T_{amb} the ambient temperature of R_G and R_L and T the temperature of the parametric amplifier, it follows that

$$F = 1 + \frac{T_{amb}}{T}\left|\frac{2Z_0}{Z - Z_0}\right|^2 \frac{R_S + R_D + \{4\omega_i^2 C_1^2 (R_i + R_D)\}^{-1}}{R_G} \tag{E.13}$$

where $Z = R_S + R_D - R_N$.

Appendix F
Noise sources in devices

The theory and a detailed description of electrical noise can be found in the related specialist literature, e.g. in the comprehensive book of Buckingham (1983) or in a collection of related publications edited by Gupta (1977). Here only a brief introduction of the most important physical noise sources involved is given.

In common circuit analysis all noise sources have to be given in the frequency domain, whereas the derivation in general is performed in the time domain. Therefore a Fourier transform has to be applied to transfer those results into the frequency domain.

F.1 THERMAL NOISE

The most common noise is the thermal Johnson noise introduced in section 3.1. It results from fluctuations in the carrier density under thermal equilibrium conditions. Because of the equipartition of the thermal energy the carriers are coupled to the thermal lattice vibrations leading to net charge variations in space and time. As a result, any dissipative system exhibits noise under thermal equilibrium conditions resulting in the thermal noise equivalent circuits shown in Fig. F.1(a), (b). Because y_n, i_n are effective values, no arrows are indicated contrary to the sources in Fig. 1.2. (Only for the evaluation of correlations are direction-arrows useful.)

F.2 SHOT NOISE

Because of the quantized electronic charge (q), shot noise after Schottky occurs if an electric current I flows under nonthermal equilibrium conditions. Its mean value is

$$I = (vq)\frac{\bar{N}}{\Delta t} \tag{F.1}$$

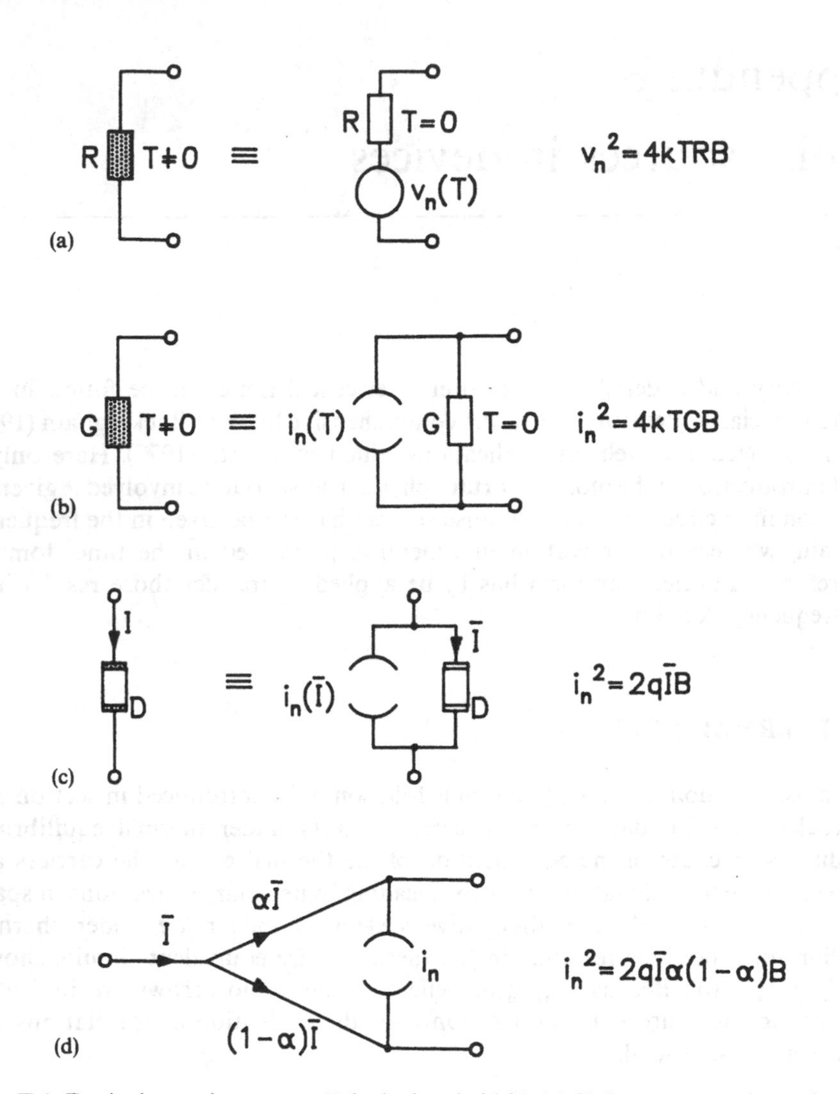

Fig. F.1 Equivalent noise sources (B is the bandwidth). (a). Voltage source, thermal noise; (b) current source, thermal noise; (c) shot noise source (D is the drift region); (d) partition noise source.

if in the time interval Δt the mean number $\bar{N}$ of charges (νq) are passing. If electrons or holes are considered, the charge factor is $\nu = 1$; in the case of ions it is, e.g. $\nu = 2$. The deviation of the current is

$$\delta I = I - \bar{I} = (nq)\frac{N - \bar{N}}{\Delta t} \tag{F.2}$$

if in a given moment in the time interval Δt the number $N \neq \bar{N}$ is passing. This leads to

$$\overline{(\delta I)^2} = (\overline{I - \bar{I}^2}) = \left(\frac{vq}{\Delta t}\right)^2 \overline{(N - \bar{N})^2} \tag{F.3}$$

With the assumption of a Laplace distribution of the number of particles involved,

$$W(N) = \frac{1}{(2\pi\bar{N})^{1/2}} \exp\left(-\frac{(N - \bar{N})^2}{2\bar{N}}\right) \tag{F.4}$$

so that

$$\overline{(N - \bar{N})^2} = \bar{N} \tag{F.5}$$

ıd therefore

$$\overline{(\delta I)^2} = \left(\frac{vq}{\Delta t}\right)^2 \bar{N} = \frac{vq}{\Delta t}\bar{I} \tag{F.6}$$

To derive the corresponding spectral noise density in the frequency domain,

$$\mathrm{d}(i_\mathrm{n})^2 = S\,\mathrm{d}f \tag{F.7}$$

the Fourier transform has to be applied. Neglecting any frequency dependence, the general convolution form is

$$S = S_\mathrm{o} = 2\Delta t \overline{F(t)^2} \tag{F.8}$$

with the mean square deviation $\overline{F(t)^2}$ in the time interval Δt (e.g. Buckingham, 1983; Burgess, 1965). Therefore, setting $\overline{F(t)^2} = \overline{\delta I^2}$ it follows that

$$S_\mathrm{o} = 2(vq)\bar{I} \tag{F.9}$$

This leads with $v = 1$ for single charged carriers (electrons, holes) to the shot noise expression after Schottky,

$$i_\mathrm{n}^2 = 2q\bar{I}B \tag{F.10}$$

with B bandwidth (Fig. F.1(c)). If the charges move with constant velocity v over a limited distance d, as is the case in semiconductor devices, e.g. along a space charge region in a junction at high electric fields ($F \gtrsim 10\,\mathrm{kV\,cm^{-1}}$), this quasistatic solution can be applied up to the transit frequency $f_\mathrm{t} = 1/2\pi t_\mathrm{t}$, where $t_\mathrm{t} = d/v$.

Taking into account the frequency dependence, instead of $S = S_\mathrm{o} \neq S(f)$ the spectral density becomes

$$S = 2(\nu q)\bar{I}\left\{\frac{\sin(\pi f t_t)}{\pi f t_t}\right\}^2 \qquad (F.11)$$

For sufficiently low frequencies it follows, in a first order approximation, that

$$d(i_n^2) = 2\nu q\bar{I}\left(1 - \frac{\pi^2 f^2 t_t^2}{3}\right)df \qquad (F.12)$$

Therefore for $F = f_t$ the error is about 8%.

Burst noise is an extension of shot noise with $\nu \gg 1$, where groups of carriers move in a statistical distributed manner. This (sometimes strong) effect cannot be covered here.

If space charges have to be taken into account the shot noise becomes reduced. Because of the quantized nature of the electron charge, however, shot noise has always to be considered in drift regions where no thermal equilibrium exists – even if the differential (RF) conductance might be zero (saturation region).

F.3 PARTITION NOISE

If under nonthermal equilibrium conditions a current is split into two parts, e.g. the emitter current I_e of a bipolar transistor into a base current I_b (by recombination) and a collector current I_c, then partition noise occurs. The moving charges either flow one way or the other, leading to fluctuations of I_b and I_c simultaneously which are fully correlated.

If

$$\alpha = \frac{\overline{I_c}}{\overline{I_e}} \qquad (F.13)$$

and

$$1 - \alpha = \frac{\overline{I_b}}{\overline{I_e}} \qquad (F.14)$$

are the mean distribution factors (quasistatic current amplification factor α in the case of a bipolar transistor, $I_b + I_c = I_e$) then a shot noise current source exists between the two branches with (Fig. F.1(d)).

$$i_n^2 = \alpha(1 - \alpha)2qI_eB \qquad (F.15)$$

The given dependence on the distribution factor α follows from the binomial distribution involved.

F.4 GENERATION–RECOMBINATION NOISE

If generation–recombination processes have to be considered, one has to distinguish between processes at one type of recombination centre and those processes which belong to a large number of centres with wide distributions of capture cross-sections and occupation lifetimes.

If carriers in a volume V are trapped in (deep) centres of the density N_T/V with the probability w, then the mean number of liberated carriers is $N = N_T(1 - w)$.

To find the mean quadratic deviation $\overline{(\delta N)^2} = \overline{(N - \bar{N})^2}$ in a given time interval Δt, the binominal distribution has to be applied leading to

$$\overline{(\delta N)^2} = N_T w(1 - w) \tag{F.16}$$

corresponding to the distribution in case of the partition noise, (section F.3). This results, in the frequency domain, to a spectral noise density $S(f)$ with

$$S(f) = \frac{S_o}{1 + (2\pi f \tau)^2} \tag{F.17}$$

where τ is the relaxation time (characteristic time constant of the carrier distribution) of the process involved. The quantity S_o is proportional to the square of the current drawn (e.g. Buckingham, 1983).

Interpreting the trap density as the dopant density, N_D/V instead, of N_T/V, the current noise observable at medium temperatures (mostly below room temperature) in semiconductors shows the same behaviour as the generation–recombination noise; the release of conducting carriers and their trapping at a dopant atom is similar.

Also the contact noise shows an $\bar{I}^2$-dependence. At a narrowing of a contact pad (bottleneck) a fluctuating voltage is developed, leading to this type noise.

F.5 FLICKER NOISE

The Flicker noise with its dependence

$$\mathrm{d}(i_n)^2 = K \frac{\bar{I}^a}{f^b} \mathrm{d}f \tag{F.18}$$

where $a = 0 \ldots 2$, $b \approx 1$, is a common phenomenon in devices. In semiconductors in general no current dependence exists, $a = 0$. Surface effects, interface traps and also trapping centres in the volume contribute to this $(1/f)$-noise, which in practice extends to extremely low frequencies ($f \lesssim 10^{-6}$ Hz). This suggests that processes with relaxation times up to hours are involved, and a very broad distribution of time constants must exist (Buckingham, 1983; Gupta, 1977; Burgess, 1965). To characterize this effect, the edge frequency f_e is used where the contribution of this $(1/f)$-noise is equal to the white noise which the device under consideration generates. If this white noise is given by the noise spectral density S_o, then

$$f_e = \frac{K}{S_o} \tag{F.19}$$

follows from

$$K\frac{\mathrm{d}f}{f_e} = S_o \,\mathrm{d}f \tag{F.20}$$

Junction FETs show relatively low values of $f_e (\lesssim 100$ Hz), whereas bipolar transistors exhibit values of $f_e \approx 100$ Hz. MOSFETs can show values of $f_e > 100$ MHz. The reason is the strong contribution of centres in the dielectric on top of the channel. Also M–S devices, e.g. Schottky diodes, often exhibit extremely high f_e-values up to GHz frequences. The reasons are the nonideal interface of a technical M–S contact and deep levels in the space charge region because of nonavoidable impurities (imperfections).

F.6 PHOTONIC NOISE

In the case of photonic devices (emitters, detectors) besides electronic noise sources the influence of spontaneous variations of the photon emission (quantum noise) has to be taken into account. The reason for this noise effect follows from the quantized light energy (photons $h\nu$), similar to the shot noise because of the quantized mobile charge q.

The spontaneous emission of a black body with

$$\bar{E} = h\nu\overline{N_{h\nu}} \tag{F.21}$$

$$\overline{P_{h\nu}} = \frac{\bar{E}}{\Delta t} = h\nu\frac{\overline{N_{h\nu}}}{\Delta t}, \tag{F.22}$$

where $\overline{N_{h\nu}}$ is the mean number of emitted photons in the time interval Δt, exhibits after Einstein and Lorentz a mean quadratic variaton of

$$\overline{(\delta E)^2} = h\nu\frac{e^{h\nu/kT}}{e^{h\nu/kT} - 1}\bar{E} \tag{F.23}$$

Therefore the mean quadratic variation of N becomes

$$\overline{\delta N_{h\nu^2}} = \frac{1}{(h\nu)^2}(\overline{\delta E^2}) = \Phi\overline{N_{h\nu}} \tag{F.24}$$

with

$$\Phi = \frac{e^{h\nu/kT}}{e^{h\nu/kT} - 1} \tag{F.25}$$

This results, in the frequency domain, to a noise contribution associated with the photon flux. The variation of the emitted optical power $\overline{P_{h\nu}}$ in a time interval Δt is

$$(\delta P_{h\nu})^2 = \frac{h\nu}{\Delta t}\Phi\overline{P_{h\nu}} \tag{F.26}$$

This corresponds, in the frequency domain, to a power noise source with

$$\mathrm{d}(P_n)^2 = 2h\nu\Phi\overline{P_{h\nu}}\,\mathrm{d}f \tag{F.27}$$

For $h\nu \gg kT$ the function Φ is $\Phi \approx 1$, leading to

$$\mathrm{d}(p_n)^2 = 2h\nu\overline{P_{h\nu}}\,\mathrm{d}f \tag{F.28}$$

This noise power is an inherent property of any photon flux.

Light-emitting devices (LEDs, diode lasers) exhibit the same noise as any p–n diode. This is transferred to the optical emission via the differential efficiency.

$$\frac{\mathrm{d}P_{h\nu}}{\mathrm{d}I} = \eta_{\mathrm{diff}}\frac{h\nu}{q} \tag{F.29}$$

Because of the nonideal fluctuating quantum efficiency $\eta = N_{h\nu}/N_\eta \neq 1$, the ratio to the number of photons generated in relation to the number of electrons in a given time interval Δt, partition noise is also involved: with

$$P_{h\nu} = \frac{N_{h\nu}h\nu}{\Delta t} \tag{F.30}$$

it is

$$\delta P_{h\nu} = \frac{h\nu}{\Delta t}\delta N_{h\nu} \tag{F.31}$$

and the mean square deviation becomes

$$\overline{(\delta P_{h\nu})^2} = \left(\frac{h\nu}{\Delta t}\right)^2 \overline{(N_{h\nu} - \overline{N_{h\nu}})^2} \tag{F.32}$$

Because (section F.3)

$$\overline{(N_{h\nu} - \overline{N_{h\nu}})^2} = \overline{N_{h\nu}}(1-\eta) \tag{F.33}$$

it follows that

$$\overline{(\delta P_{h\nu})^2} = \overline{P_{h\nu}}\frac{h\nu}{\Delta t}(1-\eta) \tag{F.34}$$

This leads, in the frequency domain, to a power fluctuation of

$$\mathrm{d}(p_{\mathrm{n}})^2 = 2h\nu\overline{P_{h\nu}}(1-\eta)\,\mathrm{d}f \tag{F.35}$$

or

$$\mathrm{d}(p_{\mathrm{n}})^2 = 2\frac{(h\nu)^2}{q}\bar{I}\eta(1-\eta)\,\mathrm{d}f \tag{F.36}$$

Also in the case of photodetectors a further photonic noise source has to be considered. Because the quantum efficiency of a photodetector is $\eta \neq 1$ and underlies statistical variations, again noise equivalent to partition noise occurs. Taking into account that the resulting photocurrent I_{ph} corresponds to αI of section F.3, with $\eta \mathrel{\hat{=}} \alpha$, it follows that

$$\mathrm{d}(i_{\mathrm{n}\eta})^2 = 2q\overline{I_{\mathrm{ph}}}(1-\eta)\,\mathrm{d}f \tag{F.37}$$

with $\overline{I_{\mathrm{ph}}}$ the mean photocurrent. The received quantum noise can also be transformed into an electrical noise source in the detector. Because the photocurrent is given by

$$I_{\mathrm{ph}} = \eta\frac{qN_{h\nu}}{\Delta t} = \eta q\frac{P_{h\nu}}{h\nu} \tag{F.38}$$

in the photodetector an equivalent noise contribution

$$\delta(I_{\mathrm{ph}})^2 = \eta\frac{q\Phi\overline{I_{\mathrm{ph}}}}{\Delta t} \tag{F.39}$$

exists.

In the frequency domain, therefore a second noise source with

$$d(i_{nQ})^2 = 2\eta q \Phi \overline{I_{ph}} \tag{F.40}$$

must be taken into account. The total noise density follows with

$$d(i_n)^2 = 2q\overline{I_{ph}}\{(1-\eta) + \eta\Phi\}\,df \tag{F.41}$$

At high quantum energies, $h\nu \gg kT$, it is $\Phi \approx 1$, and the total photonic detector noise can be expressed by the shot noise formula

$$d(i_n)^2 = 2q\overline{I_{ph}}\,df \tag{F.42}$$

Photodiodes are often biased in the breakdown region to amplify the generated photocurrent by avalanche effects. The generation of carrier cascades is a very noisy effect only reducible in its noise power by multiplying only one type of carrier (electrons or holes). For details the reader is referred to the literature (e.g. Buckingham, 1983), also regarding further noise effects such as burst noise, etc..

Appendix G
Special measurement techniques

This appendix gives some additional information regarding advanced measurement techniques in the frequency domain and time domain.

G.1 NOISE MEASUREMENTS

In the given literature many useful noise measurement setups are presented (Anastassiou and Strutt, 1974; Bächtold and Strutt, 1967; Bauer and Rothe, 1956; Fisher and Pfeiler, 1965; Fukui, 1966, 1981; Hecken, 1981; Rothe and Dahlke, 1956; Tsironis and Beneking, 1976; Uenohara, 1960). For microwave FETs a simple noise model is applicable (Pospieszalski, 1989), which requires only a single frequency noise parameter measurement.

Here the 'Y-method', a widely used method for the determination of the noise figure F of an active device, is described. If applied under different matching conditions the noise fourpole parameters after Rothe and Dahlke (1956) can also be evaluated. It consists of the device under test with its bias blocks and matching units for the variation of the effective generator impedance/input reflection r_s, e.g. by slotline transformers, and a calibrated attenuator at the output (Fig. G.1) (also Anastassiou and Strutt, 1974). A variable noise source in front of the system is used to switch between two generator noise temperatures, $T_{ex} + T_0$ and T_0, where the attenuator at the output is used to adjust both signals to the same value, indicated by a sensitive receiver.

With the attenuation being $Y > 1$ and $Y = 1$, respectively, the noise figure is

$$F = \frac{T_{ex}}{T_0(Y-1)} + \frac{1}{G_{AV}} \tag{G.1}$$

where Y is the attenuation having the same output noise power, and G_{AV} is the available gain of the test object,

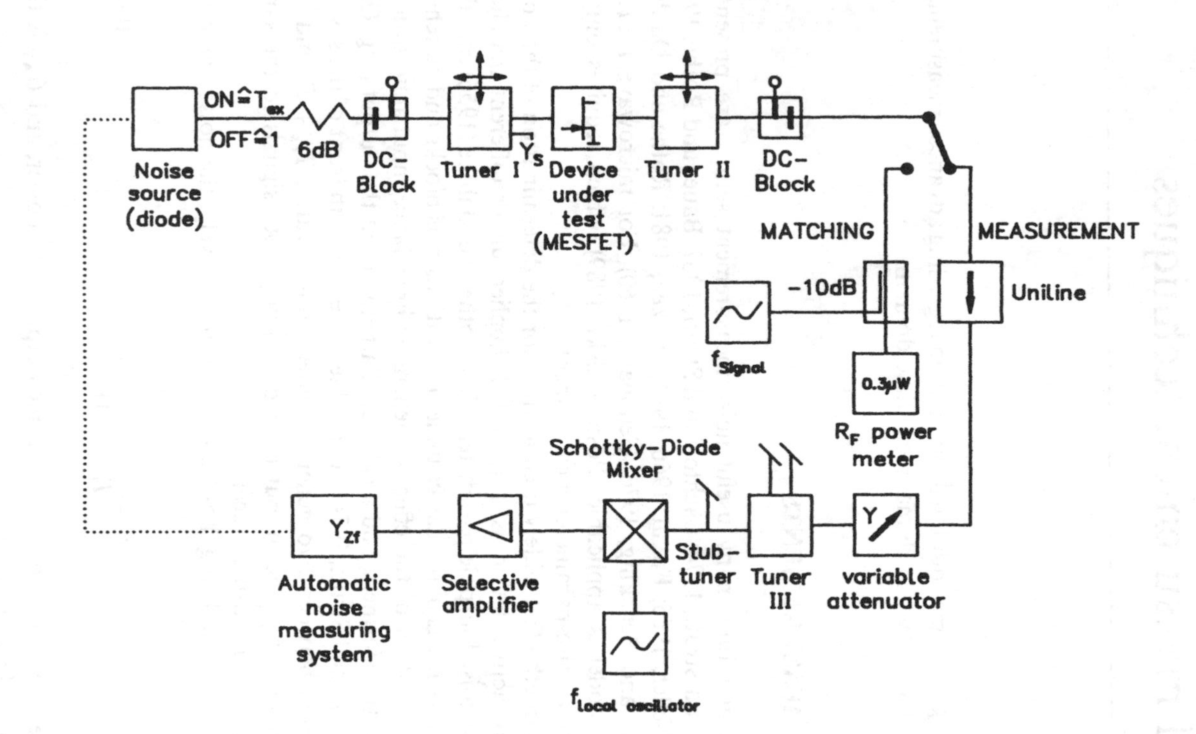

Fig. G.1 Measurement arrangement for the determination of the noise factor *F* of active devices at GHz frequencies. (After Tsironis, 1977.)

$$G_{AV}(r_s) = \frac{|S_{21}|^2(1-|r_s|^2)}{(1-|S_{22}|^2)+|r_s|^2(|S_{11}|^2-|\Delta S|^2)-2\,\mathrm{Re}\{r_s(S_{11}-S_{22}^*\Delta S)\}} \quad \text{(G.2)}$$

($\Delta S = S_{11}S_{22} - S_{22}S_{21}$) (section 2.4).

If the noise generator is switched between two positions its impedance may change. This effect must be compensated for, e.g. by using an attenuator in front of the test object.

If, e.g. 10 different generator impedances are applied, the four noise parameters (section 3.2) can be extracted. With computer programs an error minimization becomes possible.

A further problem is the characterization of the noise of oscillators. Here the two components **amplitude** noise and **phase** noise are of interest. High-Q resonators have to be used to measure the phase noise, therefore in contrast to amplifier measurements as discussed above, in general no broadband evaluation takes place.

G.2 CORRELATION MEASUREMENTS

Whereas for measurements in the frequency domain up to 30 GHz excellent systems are commercially available, this is not the case for measurements on a picosecond scale in the time domain. As Auston (1975) has shown, ultrashort electric pulses travelling on transmission lines can be generated by optically (picosecond-laser-) activated photoconductive detectors. If the lifetime of the generated carriers is short, then the device acts as a switch in the picosecond range. To measure the time response of an electrical device, the input has to be connected to a transmission line, and the output to a similar configuration, both with an optically addressable switch. If the activation of the second switch is delayed with regard to the exciting switch, then the achievable electrical signals allow a correlation measurement, and picosecond resolution becomes possible.

Figure G.2(a) shows a sketch of a coplanar version, which is very useful for these measurements because of the simple and inductance-free electrical connections to all three contacts of the electronic device (nearby and on the top side, besides the low dispersion of the coplanar line in contrast to the stripline). Furthermore the switches can be integrated into the same substrate (Si, GaAs, InP) as the device. In Fig. G.2(b) the use of the electro-optical active material lithium tantalate is shown, which reacts faster than conductive switches. However, no complete integration is possible as in the former case (Valdemanis *et al.*, 1983). Figure G.3 shows the principle of the correlation measurement and the optical part of an optoelectronic measuring setup with conductive switches (time resolution about 3 ps). The continuous variation of the time delay between the activating and measuring pulse is achieved by a rotating glass block (variable optical length). If after Mourou (in Källbäck and Beneking, 1986) electro-optical modulators

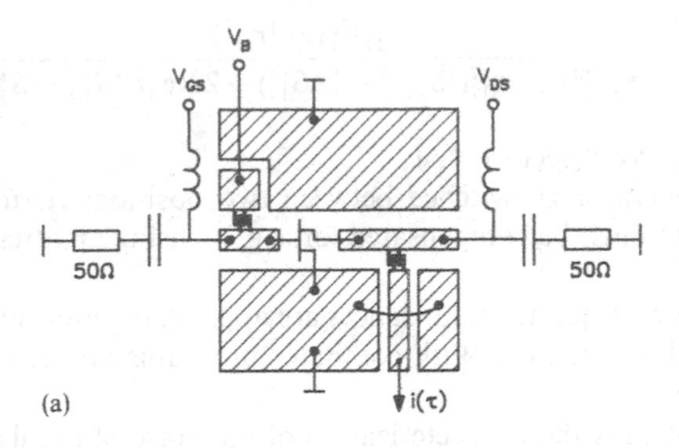

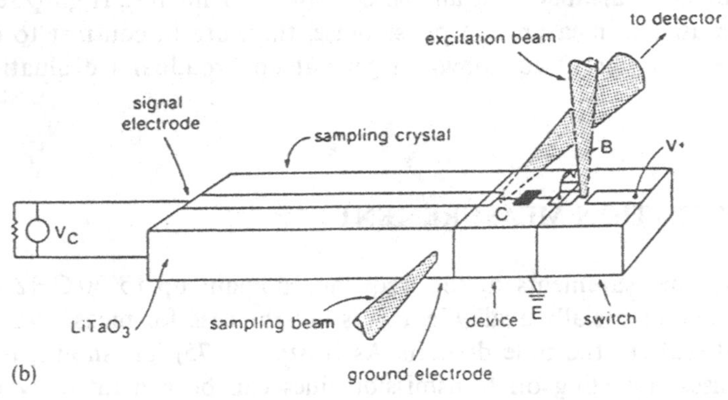

Fig. G.2 Arrangements for optoelectronic sampling. (a) Coplanar version with integrated conductive switches. (After Schumacher, 1986; Salz, 1989.) (b) use of electro-optic active material. (After Mourou, 'Picosecond electro-optic sampling', in Källbäck and Beneking, 1986, pp. 191–9.)

are used instead of photoelectric switches, even shorter pulses can be measured, down to about 0.5 ps.

Related papers can be found in Källbäck and Beneking (1986). A review summarizing different picosecond measuring techniques is given by Weingarten *et al.* (1988).

G.3 EMITTER AND COLLECTOR SERIES RESISTANCES

The parasitic resistances of bipolar transistors at the base, emitter and collector strongly influence their static and dynamic characteristics. Here the determination of the emitter and collector series resistances R_E and R_C will be treated. Their

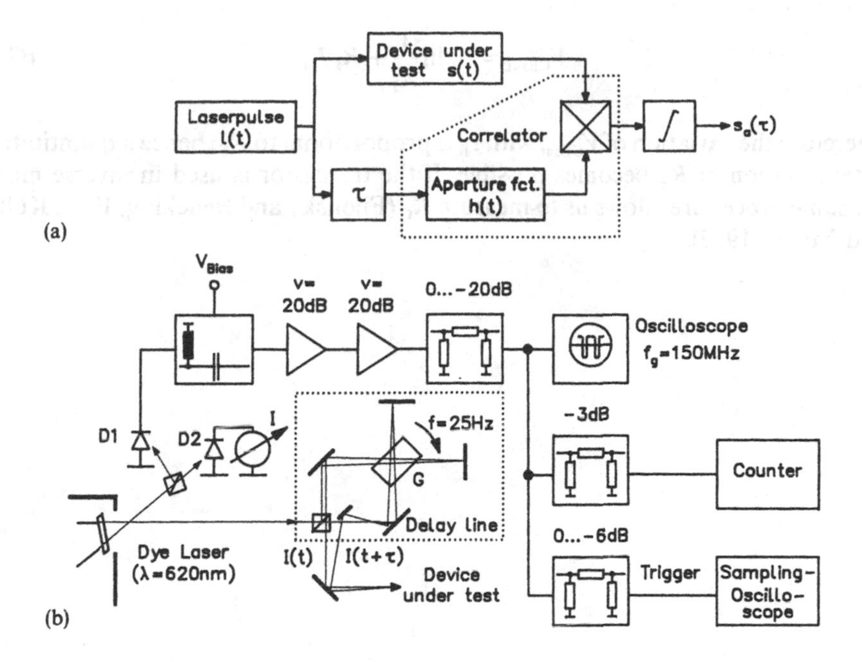

Fig. G.3 Optoelectronic correlation measurement system. (a) Principle of the set-up; (b) optical part with continuous variation of the delay time τ of the two activating optical pulses (after Salz, 1989.)

evaluation is based on the analysis of the collector–emitter saturation voltage, neglecting the small reverse saturation current,

$$V_{\mathrm{CEsat}} = V_{\mathrm{T}} \ln \frac{(I_{\mathrm{B}} B_{\mathrm{N}}/A_{\mathrm{I}}) + (I_{\mathrm{C}} B_{\mathrm{N}}/B_{\mathrm{I}})}{I_{\mathrm{B}} B_{\mathrm{N}} - I_{\mathrm{C}}} + R_{\mathrm{E}}(I_{\mathrm{B}} + I_{\mathrm{C}}) + R_{\mathrm{C}} I_{\mathrm{C}} \qquad \text{(G.3)}$$

where B_N is the common emitter current gain (normal mode), B_I the common emitter current gain (inverse mode) and A_I the common base current gain (inverse mode); positive senses of I_C, I_B leave the terminal, positive sense of I_E enters the terminal.

Even without parasitic resistances ($R_E, R_C = 0$) an offset of the origin of the I_C/V_{CE} characteristics occurs at $I_C \approx 0$. The corresponding saturation voltage reaches a limiting value (a few mV) for increasing I_B,

$$V_{\mathrm{CEsat}} = V_{\mathrm{T}} \ln \frac{1}{A_{\mathrm{I}}} \qquad \text{(G.4)}$$

With $R_E > 0$ a further displacement occurs,

$$V_{\mathrm{CEsat}} = V_{\mathrm{T}} \ln \frac{1}{A_{\mathrm{I}}} + R_{\mathrm{E}} I_{\mathrm{B}} \tag{G.5}$$

Therefore the variation of V_{CEsat} with I_{B} is proportional to R_{E}, hence a quantitative determination of R_{E} becomes possible. If the transistor is used in inverse mode the same procedure allows us to measure R_{C} (Filensky and Beneking, 1981; Kulke and Miller, 1957).

References

Actis, R. Chick, R.W., Hollis, M.A. *et al.* (1987) Small-signal gain performance of the permeable base transistor at EHF. *IEEE Electron. Dev. Lett.*, **EDL-8**, 66–8.

Anastassiou, A. and Strutt, M.J.O. (1974) Experimental and calculated four gain and four noise parameters of GaAs-FET's up to 4GHz. *Archiv El. Übertr.* **28**, 37–42.

Arai, S., Kojima, H., Otsuka, K. *et al.* (1991) U-Band 200 mW Pseudomorphic InGaAs power HEMT, in *Proceedings GaAs IC Symposium, Monterrey, Mexico*, October 20–23, pp. 105–8.

Auston, D.H. (1975) Picosecond optoelectronic switching and gating in silicon. *Appl. Phys. Lett.*, **26**, 101–6.

Bächtold, W. and Strutt, M.J.O. (1967) Darstellung der Rauschzahl und der verfügbaren Verstärkung in der Ebene des komplexen Quellenreflexionsfaktors. *Archiv Elektr. Übertragung*, **21**, 631–3.

Bauer, H. and Rothe, H. (1956) Der äquivalente Rauschvierpol als Wellenvierpol. *Archiv Elektr. Übertragung*, **10**, 241–52.

Bayraktaroglu, B. and Camilleri, N. (1988) Microwave pnp AlGaAs/GaAs heterojunction bipolar transistor. *Electron. Lett.*, **24**, 228–9.

Beneking, H. (1955) Aktive Vierpole in Zwischenbasisschaltung, *Archiv Elektr. Übertragung*, **9**, 519–27.

Beneking, H. (1959) Zur Schmalbandverstärkung mit Transistoren, *Nachrichtentechn. Zs.*, **12**, 543–6.

Beneking, H. (1965) Zum Schaltverhalten von Dioden, *Nachrichtentechnik*, **15**, 333–6.

Beneking, H. (1966) Zur Beschreibung und Kennzeichnung allgemeiner linearer Vierpole. *Archiv Elektr. Übertragung*, **20**, 254–64.

Beneking, H. III-V (1989) Semiconductor Devices, in *III-V Semiconductor Materials and Devices* (ed. R.J. Malik), Elsevier, Amsterdam, Chapter 8.

Beneking, H. and Su, L. M. (1982) Double heterojunction NPN GaAlAs/GaAs bipolar transistor. *Electron. Lett.*, **18**, 25–26.

Beneking, H., Cho. A.Y., Dekkers, J.J.M. and Morkoç H. (1982) Buried-Channel GaAs MESFET's on MBE material: scattering parameters and intermodulation signal distortion, *IEEE Trans. Electron. Dev.*, **ED-29**, 811–3.

Beneking, H., Grote, N., Roth, W. *et al.* (1980) Realization of a bipolar GaAs/GaAlAs Schottky-collector transistor. *Inst. Phys. Conf. Ser.* No. 56, 385–92.

Bosch, B.G. and Engelmann, R.W.H. (1975) *Gunn-effect Electronics*, Pitman, London.

Buckingham, M.J. (1983) *Noise in Electronic Devices and Systems*, Ellis Harwood, Chichester.

Burgess, R.E. (ed.), (1965) *Fluctuation Phenomena in Solids*, Academic Press, New York.

Carlin, H. (1956) The scattering matrix in network theory. *IRE Trans. Circuit Theory*, **CT-3**, 88–97.

Chang, C.R., Heron, P.L., and Steer, M.B. (1990) Harmonic balance and frequency-domain simulation of nonlinear microwave circuits using the block Newton method, *IEEE Trans. Microw. Theory and Techniq.*, **MTT-38**, 431–4.

Chao, P.C., Shur, M.S., Tiberio, R.C. *et al.* (1989) DC and microwave characteristics of sub-0.1-μm, gate-length planar-doped pseudomorphic HEMT's *IEEE Trans. Electron. Dev.*, **ED-36**, 461–73.

de Pian, L. (1962) *Linear Active Network Theory*, Prentice-Hall, Englewood Cliffs, NJ.

Dekkers, J.J.M., Ponse, F. and Beneking, H. (1981) Buried channel GaAs MESFET's – scattering parameter and linearity dependence on the channel doping profile. *IEEE Trans. Electron. Dev.*, **ED-28**, 1065–70.

Feldmann, J.M. (1972) *The Physics and Circuit Properties of Transistors*, John Wiley, New York, Chapter 6, p. 323 ff.

Filensky, W. and Beneking, H. (1981) New technique for determination of static emitter and collector series resistances of bipolar transistors. *Electron. Lett.*, **17**, 503–4.

Fischer, K. and Pfeiler, M. (1965) Zur Beschreibung rauschender Zweitore in Wellendarstellung. *Archiv Elektr. Übertragung*, **19**, 289–300.

Frey, J. (ed.) (1975) *Microwave Integrated Circuits*, Artech House, Dedham, MA.

Frey, J. and Bhasin, K. (eds) (1985) *Microwave Integrated Circuits*, Artech House, Dedham, MA.

Friis, H.T. (1944) Noise figure of radio receivers. *Proc. IRE*, **7**, 419–22.

Fukui, H. (1966) Available power gain, noise figure, and noise measure of two-ports and their graphical representation. *IEEE Trans. Circuit Theory*, **CT-13**, 137–42.

Fukui, H. (ed.) (1981) *Low-Noise Microwave Transistors and Amplifiers*, IEEE Press, New York.

Giacoletto, L.J. (1954) Study of P–N–P alloy junction transistor from D-C through medium frequencies. *RCA Rev.*, **15**, 506–62.

Guggenbühl, W. and Strutt, M.J.O. (1957) Theory and experiments on shot noise of semiconductor junction diodes and transistors. *Proc. IRE*, **45**, 839–54.

Gupta, M.S. (ed.) (1977) *Electric Noise: Fundamentals and Sources*, IEEE Press, New York.

Haitz, R.H. (1967) Noise of a self-sustaining avalanche discharge in silicon: low frequency noise studies. *J. Appl. Phys.*, **38**, 2935–46.

Haus, H.A. and Adler, R.B. (1959) *Circuit Theory of Linear Noisy Networks*, John Wiley, New York.

Hecken, R.P. (1981) Analysis of linear noisy two-ports using scattering waves. *IEEE Trans. Microw. Theory Techniq.* **MTT-29**, 997–1004.

Heiter, G.L. (1973) Characterization of nonlinearities in microwave devices and systems. *IEEE Trans.* **MTT-21**, 797–805.

Henke, H. (1967) Zur Höchstfrequenzverstärkung mit Tunneldioden, Diss. RWTH Aachen.

Hoffman, R.K. (1987) *Microwave Integrated Circuit Design Handbook*, Artech House, Dedham, MA.

Illi, M. (1968) Untersuchung des Schaltverhaltens von Halbleiterdioden mit Hilfe elektrischer Analogmodelle, Diss. RWTH Aachen.

Källbäck, B. and Beneking, H. (eds) (1986) *High-Speed Electronics*, part IV, *High-Speed Opto-Electronics*, Springer Series in Electronics and Photonics, Vol. 22, Springer-Verlag, Berlin.

Kingston, R.H. (1954) Switching time in junction diodes and junction transistors. *Proc. IRE*, **42**, 829–34.

Kittel, C.H. and Krömer, H. (1980) *Thermal Physics*, W.H. Freeman, San Francisco, Chapter 15.

Kulke, B. and Miller, S.L. (1957) Accurate measurement of emitter and collector series resistances in transistors. *Proc. IRE*, **45**, 90.

Kuo, F.F. (1966) *Network Analysis and Synthesis*, 2nd edn, John Wiley, New York.

Le Can, C., Hart, K. and deRuyter, C. (1962) *The Junction Transistor as a Switching Device*, Philips Techn. Bibl. Eindhoven, Netherlands.

Madihian, M., Honjo, K. Toyoshima, H. and Kumashiro, S. (1987) The design, fabrication, and characterization of a novel electrode structure self-aligned HBT with a cutoff frequency of 45 GHz. *IEEE Trans. Electron. Dev.*, **ED-34**, 1419–28.

Manley, J.M. and Rowe, H.E. (1957) Some general properties of non-linear elements – Part I. General energy relations. *Proc. IRE*, **44**, 904–13.

Mason, S.J. (1954) Power gain in feedback amplifier. *IRE Trans. Circuit Theory*, **CT-1** (2), 20–5.

Millmann, J. and Grabel, S. (1988) *Microelectronics*, 2nd Edn. part three, *Amplifier Circuit and Systems*, McGraw-Hill, New York.

Minasian, R.A. (1980) Intermodulation distortion analysis of MESFET amplifiers using the Volterra series representation. *IEEE Trans. Microw. Theory Techniq.*, **MTT-28**, 1–8.

Nyquist, H. (1928) Thermal agitation of electric charge in conductors. *Phys. Rev.*, **32**, 110–3.

Nyquist, H. (1932) Regeneration theory. *Bell Syst. Tech. J.*, **11**, 126–47.

Peng, C.K., Aksun, M.I., Ketterson, A.A. *et al.* (1987) Microwave performance of InAlAs/InGaAs/InP MODFET's *IEEE Electron. Dev. Lett.*, **EDL-8**, 24–6.

Perlow, S.M. (1976) Third-order distortion in amplifiers and mixers. *RCA Rev.*, **37**, 234–66.

Pospieszalski, M.W. (1989) Modeling of noise parameters of MESFETs and MODFETs and their frequency and temperature dependence. *IEEE Trans. Microw. Theory Techniq.*, **MTT-37**, 1340–50.

Pucel, R.A. (ed.) (1985) *Monolithic Microwave Integrated Circuits*, IEEE Press, New York.

Rauscher, C. and Tucker, R.S. (1977) Method for measuring 3rd-order intermodulation distortion in F.E.T.s *Electron. Lett.*, **13**, 701–2.

Rollett, J.M. (1962) Stability and power-gain invariants of linear twoports. *IRE Trans. Circuit Theory*, **CT-9**, 29–32.

Rollett, J.M. (1965) The measurement of transistor unilateral gain. *IEEE Trans. Circuit Theory*, **CT-12**, 91–7.

Rothe, H. and Dahlke, W. (1956) Theory of noisy fourpoles. *Proc. IRE*, **44**, 811–8.

Salz, U. (1989) Zeitbereichsmessungen im ps-Bereich mittels optisch aktivierbarer Photoleitungsschalter, Diss. RWTH Aachen.

Scanlan, J.O. (1966) *Analysis and Synthesis of Tunnel Diode Circuits*, John Wiley, London, New York.

Schumacher, H. (1986) Optoelektronische Korrelationsmeßtechnik zur Charakterisierung schneller Photodetektoren im Zeitbereich, Diss. RWTH Aachen.

Shur, M. (1987) *GaAs Devices and Circuits*, Plenum Press, New York.

Sollner, T.C.L., Goodhue, W.D., Tannenwald, P.E. *et al.* (1983) Resonant tunneling through quantum wells at frequencies up to 2.5 THz. *Appl. Phys. Lett.*, **43**, 588–90.

Sze, S.M. (ed.) (1990) *High-Speed Semiconductor Devices*, John Wiley, New York.

Tanaka, K., Ogawa, M., Togashi, K. *et al.* (1986) Low-noise HEMT using MOCVD. *IEEE Transact.*, **MTT-34**, 1522–7.

Thoma, W. (1969) Ein Beitrag zur Analyse des Großsigalverhaltens schwingkreisbelasteter Transistorstufen, Diss. RWTH Aachen.

Trew, R.J. and Steer, M.B. (1987) Millimeter-wave performance of state-of-the-art MESFET, MODFET and PBT transistors. *Electron. Lett.*, **23**, 149–51.

Tsironis, C. (1977) Untersuchungen zum Rauschverhalten des Galliumarsenid-Schottky-Gate Feldeffekttransistors im GHz-Bereich, Diss. RWTH Aachen.

Tsironis, C. and Beneking, H. (1976) Verfahren zur exakten Bestimmung der Rauschparameter aktiver Zweitore im Frequenzbereich 1 GHz bis 11 GHz. *Nachrichtentechn. Zs.*, **29**, 385–9.

Uenohara, M. (1960) Noise considerations of the variable capacitance amplifier. *Proc. IRE*, **48**, 169–79.

Unger, H.G. and Harth, W. (1972) *Hochfrequenz-Halbleiterelektronik*, S. Hirzel, Stuttgart, Chapter 1.8.

Valdemanis, J.A. Mourou, G.A. and Gabel, C.W. (1983) Subpicosecond electrical sampling. *IEEE J. Quant. Electron.*, **19**, 664–7.

Vogelsang, E. (1963) Ein Beitrag zur Kennlinie und zum Ersatzschaltbild der Tunneldiode, Diss. RWTH Aachen.

Weingarten, K.J., Rodwell, M.J.W. and Bloom, D. M. (1988) Picosecond optical sampling of GaAs integrated circuits. *IEEE Trans. Quant. Electron.*, **QE-4**, 198–220.

Solutions to exercises

CHAPTER 1

1.1 Because the h-parameters link the input voltage and output current to the input current and output voltage (Appendix A), the corresponding DC dependencies are $V_1 = V_1(I_1, V_2)$ and $I_2 = I_2(I_1, V_2)$. Therefore the two-dimensional Taylor series are (for bias conditions I_{1P}, V_{2P})

$$V_1 = V_{1P} + \frac{\partial V_1}{\partial I_1}(I_1 - I_{1P}) + \frac{\partial V_1}{\partial V_2}(V_2 - V_{2P}) + \frac{1}{2}\frac{\partial^2 V_1}{\partial I_1^2}(I_1 - I_{1P})^2 + \frac{1}{2}\frac{\partial^2 V_1}{\partial V_2^2}(V_2 - V_{2P})^2 + \frac{\partial V_1}{\partial I_1}\frac{\partial V_1}{\partial V_2}(I_1 - I_{1P})(V_2 - V_{2P}) + \cdots$$

and

$$I_2 = I_{2P} + \frac{\partial I_2}{\partial I_1}(I_1 - I_{1P}) + \frac{\partial I_2}{\partial V_2}(V_2 - V_{2P}) + \frac{1}{2}\frac{\partial^2 I_2}{\partial I_1^2}(I_1 - I_{1P})^2 + \frac{1}{2}\frac{\partial^2 I_2}{\partial V_2^2}(V_2 - V_{2P})^2 + \frac{\partial I_2}{\partial I_1}\frac{\partial I_2}{\partial V_2}(I_1 - I_{1P})(V_2 - V_{2P}) + \cdots$$

For sufficiently small amplitudes,

$$i_1 = I_1 - I_{1P}$$
$$v_1 = V_1 - V_{1P}$$
$$i_2 = I_2 - I_{2P}$$
$$v_2 = V_2 - V_{2P}$$

It follows that

$$v_1 = \left.\frac{\partial V_1}{\partial I_1}\right|_{V_2 = \text{const.}} \cdot i_1 + \left.\frac{\partial V_1}{\partial V_2}\right|_{I_1 = \text{const.}} \cdot v_2$$

$$i_2 = \left.\frac{\partial I_2}{\partial I_1}\right|_{V_2 = \text{const.}} \cdot i_1 + \left.\frac{\partial I_2}{\partial V_2}\right|_{I_1 = \text{const.}} \cdot v_2$$

These equations allow us to extract the h-parameters:

$$h_{11} = \left.\frac{\partial V_1}{\partial I_1}\right|_{V_2 = \text{const.} \triangleq v_2 = 0}$$

$$h_{12} = \left.\frac{\partial V_1}{\partial V_2}\right|_{I_1 = \text{const.} \triangleq i_1 = 0}$$

$$h_{21} = \left.\frac{\partial I_2}{\partial I_1}\right|_{V_2 = \text{const.} \triangleq v_2 = 0}$$

$$h_{22} = \left.\frac{\partial I_1}{\partial V_2}\right|_{I_1 = \text{const.} \triangleq i_1 = 0}$$

As can be seen,

- $h_{11} = h_i$ is the input resistance for short-circuited output;
- $h_{12} = h_r$ represents the inner feedback voltage at open input;
- $h_{21} = h_m$ is the current gain for short-circuited output;
- $h_{22} = h_o$ is the output conductance for open input.

At higher frequencies these quantities become complex, as shown for the y-parameters in section 1.1.

The quantities h_{12} and h_{21} have no dimensions, they are (voltage and current) ratios. The parameter h_{11} corresponds to an impedance, whereas h_{22} represents an admittance. That is the reason for their nomination 'hybrid' parameters (however, the p-parameters are also hybrid).

1.2 The h-parameters form the matrix

$$v_1 = h_{11} i_1 + h_{12} v_2$$
$$i_2 = h_{21} i_1 + h_{22} v_2$$

and the y-parameters

$$i_1 = y_{11} v_1 + y_{12} v_2$$
$$i_2 = y_{21} v_1 + y_{22} v_2$$

To evaluate y_{11} we have to determine

$$\left.\frac{i_1}{v_1}\right|_{v_2 = 0}$$

in the h-matrix configuration. From the first row it follows directly that

$$\frac{1}{y_{11}} = h_{11} \quad \text{or} \quad y_{11} = \frac{1}{h_{11}}$$

The reverse mutual conductance y_{12} is

$$y_{12} = \left.\frac{i_1}{v_2}\right|_{v_1=0}$$

Therefore, with

$$-h_{11}i_1 = h_{12}v_2$$

from the first row it follows that

$$y_{12} = -\frac{h_{12}}{h_{11}}$$

The transconductance

$$y_{21} = \left.\frac{i_2}{v_1}\right|_{v_2=0}$$

follows from the second row with

$$y_{21} = h_{21}\left.\frac{i_1}{v_1}\right|_{v_2=0} = \frac{h_{21}}{h_{11}}$$

and the output admittance

$$y_{22} = \left.\frac{i_2}{v_2}\right|_{v_1=0}$$

becomes

$$h_{22} + h_{21}\left.\frac{i_1}{v_2}\right|_{v_1=0} = h_{22} + h_{21}y_{12}$$

or

$$y_{22} = h_{22} - \frac{h_{12}h_{21}}{h_{11}}$$

The dual procedure is used to evaluate the h-parameters by the y-parameters, leading to

$$h_{11} = \frac{1}{y_{11}}$$

$$h_{12} = -\frac{y_{12}}{y_{11}}$$

$$h_{21} = \frac{y_{21}}{y_{11}}$$

$$h_{22} = y_{22} - \frac{y_{12}y_{21}}{y_{11}}$$

1.3 With

$$I_1 = I_{1P} + \frac{\partial I_1}{\partial V_1}(V_1 - V_{1P}) + \frac{\partial I_1}{\partial V_2}(V_2 - V_{2P}) + \frac{\partial I_1}{\partial V_3}(V_3 - V_{3P}) + \cdots$$

$$I_2 = I_{2P} + \frac{\partial I_2}{\partial V_1}(V_1 - V_{1P}) + \frac{\partial I_2}{\partial V_2}(V_2 - V_{2P}) + \frac{\partial I_2}{\partial V_3}(V_3 - V_{3P}) + \cdots$$

$$I_3 = I_{3P} + \frac{\partial I_3}{\partial V_1}(V_1 - V_{1P}) + \frac{\partial I_3}{\partial V_2}(V_2 - V_{2P}) + \frac{\partial I_3}{\partial V_3}(V_3 - V_{3P}) + \cdots$$

we have

$$i_1 = g_{11}v_1 + g_{12}v_2 + g_{13}v_3$$
$$i_2 = g_{21}v_1 + g_{22}v_2 + g_{23}v_3$$
$$i_3 = g_{31}v_1 + g_{32}v_2 + g_{33}v_3$$

Therefore

$$g_{11} = \left.\frac{\partial I_1}{\partial V_1}\right|_{v_2,v_3=0}, \quad g_{12} = \left.\frac{\partial I_1}{\partial V_2}\right|_{v_1,v_3=0}, \quad g_{13} = \left.\frac{\partial I_1}{\partial V_3}\right|_{v_1,v_2=0}$$

$$g_{21} = \left.\frac{\partial I_2}{\partial V_1}\right|_{v_2,v_3=0}, \quad g_{22} = \left.\frac{\partial I_2}{\partial V_2}\right|_{v_1,v_3=0}, \quad g_{23} = \left.\frac{\partial I_2}{\partial V_3}\right|_{v_1,v_2=0}$$

$$g_{31} = \left.\frac{\partial I_3}{\partial V_1}\right|_{v_2,v_3=0}, \quad g_{32} = \left.\frac{\partial I_3}{\partial V_2}\right|_{v_1,v_3=0}, \quad g_{33} = \left.\frac{\partial I_3}{\partial V_3}\right|_{v_1,v_2=0}$$

All these parameters have to be measured under the condition that the other ports are dynamically short-circuited (besides the established bias condition). As in the case of the fourpole parameters the complex y-parameters y_{kl} are the extension of the differential parameters g_{kl} derived above, $y_{kl} = g_{kl} + jb_{kl}$.

1.4 To introduce the z-parameters into the formula

$$S_{11} = r_1\bigg|_{r_L=0} = \frac{Z_1 - Z_0}{Z_1 + Z_0}$$

(where Z_0 is characteristic impedance of the transmission line at the input, Z_1 the input impedance of the twoport) the relevant impedances at the input

$$Z_1 = z_{11} - \frac{z_{12}z_{21}}{z_{22} + Z_L}$$

and at the output

$$Z_2 = z_{22} - \frac{z_{12}z_{21}}{z_{11} + Z_G}$$

have to be implemented ($Z_G = Z_0$ is the generator impedance, $Z_L = Z_2$ for $r_L = 0$).

Inserting these expressions in the formula for S_{11}, it follows that

$$S_{11} = \frac{z_{11} - \dfrac{z_{12}z_{21}}{z_{22} + \left(z_{22} - \dfrac{z_{12}z_{21}}{z_{11} + Z_G}\right)} - Z_G}{z_{11} - \dfrac{z_{12}z_{21}}{z_{22} + \left(z_{22} - \dfrac{z_{12}z_{21}}{z_{11} + Z_G}\right)} + Z_G}$$

which can be written as

$$S_{11} = 1 - \frac{Z_G z_{22}}{\Delta_z + Z_G z_{22}}$$

where $\Delta_z = z_{11}z_{22} - z_{12}z_{21}$.

For S_{21} we have

$$S_{21} = \frac{2z_{21}(Z_G Z_L)^{-1/2}}{1 + \dfrac{z_{11}}{Z_G} + \dfrac{z_{22}}{Z_L} - \Delta_z}$$

1.5 For the evaluation we refer to Appendix B, where the wave source is introduced. First voltages and currents have to be implemented into the wave source equation: with

$$b_G = r_G a_G + b_Q$$

and

$$a = \tfrac{1}{2}(v + i)$$

$$b = \tfrac{1}{2}(v - i)$$

where

$$v = \frac{V}{(Z_0)^{1/2}}, \quad i = I(Z_0)^{1/2}$$

and

$$a = \frac{V^+}{(Z_0)^{1/2}} = I^+(Z_0)^{1/2}, \quad b = \frac{V^-}{(Z_0)^{1/2}} = I^-(Z_0)^{1/2}$$

we have

$$b_G = \tfrac{1}{2}(v_G - i_G) = \tfrac{1}{2}(v_G + i_G)r_G + b_Q$$

or

$$V_G = v_G(Z_0)^{1/2} = I_G Z_0 \frac{1 + r_G}{1 - r_G} + \frac{2b_Q(Z_0)^{1/2}}{1 - r_G}$$

With the impedance Z_G of the generator,

$$r_G = \frac{Z_G - Z_0}{Z_G + Z_0}$$

and therefore

$$Z_0 = Z_G \frac{1 - r_G}{1 + r_G}$$

The voltage at the source port becomes

$$V_G = I_G Z_G + \frac{2b_Q(Z_0)^{1/2}}{1 - r_G}$$

Because here the positive I-direction is directed into the port (Fig. B.1(c)), this can be interpreted as a conventional voltage generator with the inner voltage (Fig. B.1(d))

$$V_{G0} = \frac{2b_Q(Z_0)^{1/2}}{1 - r_G}$$

The available power of this equivalent generator is

$$P_{AV,G} = \frac{|V_{G0}|^2}{8R_G}$$

where

$$R_G = \text{Re}\{Z_G\}$$

In terms of b_Q this is

$$P_{AV,G} = \frac{4Z_0|b_Q|^2}{|1 - r_G|^2} \frac{1}{8R_G}$$

which should be equal to

$$P_{1\max} = \frac{|b_Q|^2}{2} \frac{1}{|1 - r_G|^2}$$

If so, then it should be

$$Z_0(1 - |r_G|^2) = R_G|1 - r_G|^2$$

With

$$r_G = \frac{Z_G - Z_0}{Z_G + Z_0}$$

this is

$$Z_0\left(1 - \left|\frac{Z_G - Z_0}{Z_G + Z_0}\right|^2\right) = R_G\left|1 - \frac{Z_G - Z_0}{Z_G + Z_0}\right|^2$$

or

$$Z_0\{|Z_G + Z_0|^2 - |Z_G - Z_0|^2\} = R_G(2Z_0)^2$$

Because

$$\{|Z_G + Z_0|^2 - |Z_G - Z_0|^2\} = 4R_GZ_0$$

both sides of the equation are equal, and

$$P_{1\max} = P_{AVG}$$

as claimed above.

1.6 Referring to Fig. Ex.1.1, and from section 1.2.1,

$$b_1^{\mathrm{I}} = T_{11}^{\mathrm{I}} a_2^{\mathrm{I}} + T_{12}^{\mathrm{I}} b_2^{\mathrm{I}}$$
$$a_1^{\mathrm{I}} = T_{21}^{\mathrm{I}} a_2^{\mathrm{I}} + T_{22}^{\mathrm{I}} b_2^{\mathrm{I}}$$

or

$$\boldsymbol{W}_1^{\mathrm{I}} = \boldsymbol{T}^{\mathrm{I}} \cdot \boldsymbol{W}_2^{\mathrm{I}}$$

with

$$\boldsymbol{W}_1^{\mathrm{I}} = \begin{bmatrix} b_1^{\mathrm{I}} \\ a_1^{\mathrm{I}} \end{bmatrix} \quad \boldsymbol{W}_2^{\mathrm{I}} = \begin{bmatrix} a_2^{\mathrm{I}} \\ b_2^{\mathrm{I}} \end{bmatrix}$$

Analogously,

$$\boldsymbol{W}_1^{\mathrm{II}} = \boldsymbol{T}^{\mathrm{II}} \cdot \boldsymbol{W}_2^{\mathrm{II}}$$

with

$$\boldsymbol{W}_1^{\mathrm{II}} = \begin{bmatrix} b_1^{\mathrm{II}} \\ a_1^{\mathrm{II}} \end{bmatrix} \quad \boldsymbol{W}_2^{\mathrm{II}} = \begin{bmatrix} a_2^{\mathrm{II}} \\ b_2^{\mathrm{II}} \end{bmatrix}$$

Because

$$\boldsymbol{W}_2^{\mathrm{I}} = \boldsymbol{W}_1^{\mathrm{II}}$$

it follows that

$$\boldsymbol{W}_1^{\mathrm{I}} = \boldsymbol{T}^{\mathrm{I}} \cdot (\boldsymbol{T}^{\mathrm{II}} \cdot \boldsymbol{W}_2^{\mathrm{II}})$$

Therefore the cascaded twoports are described by

$$\boldsymbol{W}_1^{\mathrm{I}} = \boldsymbol{T}_{\Sigma} \cdot \boldsymbol{W}_2^{\mathrm{II}}$$

where

$$\boldsymbol{T}_{\Sigma} = \boldsymbol{T}^{\mathrm{I}} \cdot \boldsymbol{T}^{\mathrm{II}}$$

1.7 Because a lossless twoport exhibits an input reflection factor

$$r_1 = \frac{S_{11}}{S_{22}^*} \frac{S_{22}^* - r_{\mathrm{L}}}{1 - S_{22} r_{\mathrm{L}}}$$

and a twoport, which is additionally reciprocal ($S_{12} = S_{21}$),

$$r_1 = \frac{S_{22}}{S_{11}^*} \frac{S_{22}^* - r_{\mathrm{L}}}{1 - S_{22} r_{\mathrm{L}}}$$

it follows that

$$\frac{S_{11}}{S_{22}^*}=\frac{S_{22}}{S_{11}^*} \quad \text{or} \quad |S_{11}|=|S_{22}|$$

To achieve $r_1=0$ we must have

$$S_{22}^*=r_L \quad \text{or} \quad S_{22}=0.6e^{-j3\pi/4}$$

With the given phase of S_{11}, arc $S_{11}=\pi/4$, it follows, with $|S_{11}|=|S_{22}|$, that

$$S_{11}=0.6e^{-j3\pi/4}$$

The general expression for r_1,

$$r_1=S_{11}+\frac{S_{12}S_{21}r_L}{1-S_{22}r_L}$$

leads, for $r_1=0$ (because $S_{12}=S_{21}$, reciprocal twoport) to

$$0=S_{11}+\frac{S_{12}^2r_L}{1-S_{22}r_L}$$

or

$$-S_{11}(1-S_{22}r_L)=S_{12}^2r_L$$

This leads, with $|S_{11}|=|S_{22}|=0.6$ and the given phases $\varphi_{11}=\pi/4$ and $\varphi_{22}=-3\pi/4$, to

$$S_{12}^2=S_{11}\left(S_{22}-\frac{1}{r_L}\right)=0.6e^{j\pi/4}\left(0.6e^{-j3\pi/4}-\frac{1}{0.6}e^{-j3\pi/4}\right)$$
$$=-0.64e^{-j\pi/2}=0.64e^{j\pi/2}$$

Therefore

$$S_{12}=\pm(0.64e^{j\pi/2})^{1/2}=0.8\cdot e^{\pm j\pi/4}$$

equal to S_{21}.

1.8 In section 1.2.1 the scattering matrix was introduced. In the case of a threeport it is

$$b_1=S_{11}a_1+S_{12}a_2+S_{13}a_3$$
$$b_2=S_{21}a_1+S_{22}a_2+S_{23}a_3$$
$$b_3=S_{31}a_1+S_{32}a_2+S_{33}a_3$$

where b are outgoing waves, a incoming waves.

To determine the parameters

$$|\rho| = |S_{11}| = |S_{22}| = |S_{33}|$$
$$|\sigma| = |S_{12}| = |S_{23}| = |S_{31}|$$
$$|\tau| = |S_{13}| = |S_{21}| = |S_{32}|$$

the circulator is connected to a generator at port 1 and to matched loads at ports 2 and 3, respectively (Fig. Ex1.2). Therefore

$$|a_2| = |a_3| = 0 \quad \text{and} \quad \left|\frac{b_1}{a_1}\right| = |S_{11}| = |\rho|$$

correspond to the entrance port reflection

$$\frac{P_1}{P_0} = \left|\frac{b_1}{a_1}\right|^2 = |\rho|^2$$

The quantity $|\tau|$ follows, with

$$\frac{P_2}{P_0} = \left|\frac{b_2}{a_1}\right|^2 = |\tau|^2$$

and for σ

$$\frac{P_3}{P_0} = \left|\frac{b_3}{a_1}\right|^2 = |\sigma|^2$$

With $P_3/P_0 \triangleq -20\,\text{dB}$,

$$|\sigma|^2 = \frac{1}{100} \text{ or } |\sigma| = 0.1$$

Because $P_2/P_0 \triangleq -3\,\text{dB}$, $|\tau| = 2^{1/2}$

Finally, with the given input reflection of $-26\,\text{dB}$ it follows that

$$|\rho|^2 = 0.0025 \quad \text{or} \quad |\rho| = 0.05$$

CHAPTER 2

2.1 Show first that the voltage gain G_V and the current gain G_I are expressed by the given relationships, following from both equivalent circuits at the output of the fourpole. Identify G_V by the p-parameters and G_I by the h-parameters.

Combine the two expressions to achieve the power gain $G_{eff}(a_2, f)$ and follow the advice in Exercise 2.1. Then draw the dependence of the normalized gain $g_{eff} = G_{eff}/G_0$ depending on a_2 (for $a_2 < 1$, linear scale) and $1/a_2$ (for $a_2 > 1$, inverse scale), respectively, for the given parameters f.

By combining g_{eff} with the input matching (a_1) the three-dimensional graphical representation of the normalized transducer gain g_T can be drawn. The result is given in Fig. E2.2(b), which together with the curves in Fig. 2.2(a) shows the influence of the matching on both sides of the active fourpole. To achieve absolute values for $f = 1$ compute $g_T(a_1, a_2)$ for $a_1, a_2 = 0.5$; $a_1, a_2 = 2.0$ and compare this result with the maximum value for $a_1, a_2 = 1$ (leading to 0.79 and 1.0, respectively).

2.2 As long as both parameters are equal, $y_{22} = h_{22}$, no reverse mutual influence exists. This follows from the two expressions

$$y_{22} = h_{22}\left(1 - \frac{h_{12}h_{21}}{h_{11}h_{22}}\right) \quad \text{and} \quad h_{22} = y_{22}\left(\frac{1 - y_{12}y_{21}}{y_{11}y_{22}}\right)$$

where the quantities

$$\delta_H = \frac{h_{12}h_{21}}{h_{11}h_{22}} \quad \text{and} \quad \delta_Y = \frac{y_{12}y_{21}}{y_{11}y_{22}}$$

determine with

$$1 - \delta_Y = \frac{1}{1 - \delta_H} = \frac{h_{22}}{y_{22}} = f^2$$

the ratio of the two output conductances (real parameters).

From the results of Exercise 2.1 it follows for the optimum transducer gain that

$$G_{Tmax} = G_0 \frac{4}{(1 + f)^2}$$

The value $f = 1$ corresponds to a fourpole without feedback, $h_{22} = y_{22}$. If $f > 1$ the gain is reduced and negative feedback is implemented. Contrary, for $f < 1$, the gain is enhanced, therefore positive feedback occurs, and the fourpole becomes more unstable (results of Exercise 2.1). The stability limits are $\delta_Y = 1$ and $\delta_H = 1$, respectively, where y_{22} or h_{22} become zero.

2.3 To simplify the evaluation a normalized version of the parameters will be used with $\Omega = \omega/\omega_0$. Introducing this quantity it follows that

$$y_{11} = 0.1\,\text{mS} + \text{j}\Omega\cdot 0.4\,\text{mS}$$
$$y_{12} = -\text{j}\Omega\cdot 0.2\,\text{mS}$$
$$y_{21} = 16\,\text{mS}(1+\text{j}\Omega)^{-1}$$
$$y_{22} = 2\,\text{mS}$$

The stability constant $\boldsymbol{d} = \delta + \text{j}\Delta$ can therefore be expressed as

$$\boldsymbol{d} = 16\frac{-\text{j}\Omega}{1+\text{j}\Omega}$$

with

$$\delta = 16\frac{-\Omega^2}{1+\Omega^2}$$

and

$$\Delta = 16\frac{-\Omega}{1+\Omega^2}$$

Therefore the frequency dependence of $\boldsymbol{d}$ can be drawn (Fig. Ex2.3).

To determine f_{crit}, the value of $\Omega = \Omega_{\text{crit}}$ for $\delta + (\Delta^2/4) = 1$ has to be computed. Than $f_{\text{crit}} = \Omega_{\text{crit}}(\omega_0/2\pi)$ can be evaluated. It follows that $f_{\text{crit}} = 29.6\,\text{MHz}$. A second value follows from the computation, $f_{\text{crit2}} =$ 328 MHz, at which frequency the transistor becomes again absolutely stable; the curve $\boldsymbol{d}\,(\Omega)$ crosses the curve of the stability limit twice.

If the shunt resistor $R = 40\,\text{k}\Omega$ is applied, the varied curve d starts for $\Omega = 0$ at $\delta = -2$ instead of $\delta = 0$, and it follows that with the modified value $y_{12} = 25\,\mu\text{s} - \text{j}\Omega\cdot 0.2\,\text{mS}$,

$$\boldsymbol{d} = -16\frac{\frac{1}{8}+\text{h}\Omega}{1+\text{j}\Omega}$$

Real and imaginary parts become

$$\delta = -\frac{2+16\Omega^2}{1+\Omega^2}$$

and

$$\Delta = -\frac{14\Omega}{1+\Omega^2}$$

This leads to the modified critical frequencies

$$f_{crit} = 66.5\,\text{MHz}$$

and

$$f_{crit2} = 253\,\text{MHz}$$

Therefore the critical frequency range $\Delta f = f_{crit2} - f_{crit}$ is reduced, where instability could occur. With the values for f_{crit} the corresponding MSG values can be computed, whereas the low frequency values of the gain follows simply via G_{T_O} (section 2.2).

2.4 In Fig. Ex2.4 the circuit of the narrowband amplifier is shown, and also the equivalent circuit, related to the output of the first stage. The output power is

$$P_O = |V_O|^2 G_O$$

and its maximum value, obtained for impedance matching

$$G_P + G_O = G_i$$

becomes

$$P_{Omax} = \frac{|I_G|^2}{4G_i}$$

With

$$G = G_i + G_P + G_O$$

and the characteristic impedance of the resonant circuit

$$Z = \omega_0 L = \left(\frac{L}{C + C_i + C_O}\right)^{1/2} = \frac{1}{G_Q}$$

($f_0 = \omega_0/2\pi$ is the resonance frequency and Q is the quality factor of the complete resonant circuit) it follows that at resonance,

$$\omega = \omega_0 = \frac{1}{(L(C + C_i + C_O))^{1/2}}$$

for the voltage

$$V_O = \frac{I_O}{G}$$

The ratio of the relevant power P with respect to P_{max} becomes

$$\eta = \frac{P_0}{P_{0max}} = \frac{4G_iG_0}{G^2}$$

Setting

$$\frac{Q}{Q_0} = \frac{G_P}{G} = q$$

($q \leqslant 1$; Q_0 is the quality factor of the connected resonance circuit alone) and

$$\frac{G_i}{G_P} = g$$

it follows that

$$\eta = 4gq\{1 - q(1+g)\}$$

With this equation the optimum value of the load conductance G_0 for a given value of Q can be determined. By differentiation with $q = \text{const.}$ it is

$$g_{opt} = \frac{1-q}{2q} \quad \text{or} \quad G_{0opt} = G_i$$

With

$$\eta_{opt} = (1-q)^2$$

the optimum power in the load becomes

$$P_{opt} = P_{0max}\left(1 - \frac{Q}{Q_0}\right)^2$$

For $Q \approx Q_0$ a strong reduction in the gain results; only for $Q_0 \gg Q$ does $P_{opt} \approx P_{0max}$.

2.5 To determine the electrical behaviour of the combination of the lossless transmission line (length l, phase angle φ) in front of the twoport its transmission matrix ($\boldsymbol{T}_1$) has to be combined (multiplied) with that of the twoport ($\boldsymbol{T}_{II}$). To verify that the lossless transmission line does not change the transmitted power it must be shown that input power P_{1I} and output power $P_{2I} = P_{1II}$ of the corresponding twoport ($\boldsymbol{T}_I$) are equal.

With the transmission matrix

$$\boldsymbol{T}_I = \begin{bmatrix} e^{-j\varphi} & 0 \\ 0 & e^{j\varphi} \end{bmatrix}$$

first the input power

$$P_{11} = \frac{a_{11}a_{11}^*}{2} - \frac{b_{11}b_{11}^*}{2}$$

has to be evaluated (section 1.2).

With the general expressions

$$b_1 = T_{11}a_2 + T_{12}b_2$$
$$a_1 = T_{21}a_2 + T_{22}b_2$$

it follows that

$$P_{11} = \frac{|b_{21}|^2}{2} - \frac{|a_{21}|^2}{2}$$

because

$$b_{11} = e^{-j\varphi}a_{21}, \quad a_{11} = e^{j\varphi}b_{21}$$

The output power P_{21} is identical with the input power of the second twoport, $P_{21} = P_{1II}$. Because $a_{1II} = b_{21}$, $b_{1II} = a_{21}$, the input power P_{11} can be expressed by

$$P_{11} = \frac{|a_{1II}|^2}{2} - \frac{|b_{1II}|^2}{2}$$

which is equal to $P_{1II} \equiv P_{21}$.

Therefore the power gain of the second twoport is not changed if the lossless transmission line is connected to the twoport (being evident because no lossy circuit elements are incorporated).

2.6 This problem is similar to Exercise 2.5. The $\boldsymbol{T}$-parameters of a twoport are given by (section 1.2)

$$T_{11} = \left.\frac{b_1}{a_2}\right|_{b_2=0}$$

$$T_{12} = \left.\frac{b_1}{b_2}\right|_{a_2=0}$$

$$T_{21} = \left.\frac{a_1}{a_2}\right|_{b_2=0}$$

$$T_{22} = \left.\frac{a_1}{b_2}\right|_{a_2=0}$$

In the case of the lossless transmission line of length l only a phase delay φ is effective, where

$$\varphi = \beta_1 = \frac{2\pi l}{\lambda}$$

The wave b_1 becomes

$$b_1 = a_2 e^{-j\varphi}$$

leading to

$$T_{11} = e^{-j\varphi}$$

Similarly,

$$T_{22} = e^{+j\varphi}$$

whereas the other two T parameters are zero.

Combining this twoport with the generator, the transmission line with its characteristic impedance Z_0 acts as a transformer, and the inner resistance R_G of the generator becomes transformed, leading to the impedance

$$R_G' = \frac{R_G + jZ_0 \tan(\beta_l)}{1 + j\left(\frac{R_G}{Z_0}\right)\tan(\beta_l)}.$$

at the output port of the transmission line.

The voltage also becomes transformed. Because the available power has to be the same, it follows that with

$$P_{max} = \frac{V_G^2}{4R_G} = \frac{(V'_G)^2}{4R'_G}$$

the expression for the voltage V'_G is

$$V'_G = V_G\left(\frac{R'_G}{R_G}\right)^{1/2}$$

The resulting description in the Smith chart can be drawn in Fig. E2.5 where, depending on the length l, a left-turning circle results. For $l = 0$, $R'_G/Z_0 = 0.5$, and for $l = \lambda/4$, $R'_G/Z_0 = 2$. For $l = \lambda/2$ again $R'_G/Z_0 = 0.5$ is reached etc.

2.7 The equivalence of the given expressions follows by the application of the parameter transformation listed in Appendix A. Because

$$h_{21} = \frac{y_{21}}{y_{11}}, \quad h_{12} = \frac{-y_{12}}{y_{11}}$$

we have

$$\left|\frac{y_{21}}{y_{12}}\right| = \left|\frac{h_{21}}{h_{12}}\right|$$

and by using the transformation formula for the conversion of the y-parameter into the S-parameters it follows that

$$\left(\frac{S_{21}}{S_{12}}\right) = \left|\frac{y_{21}}{y_{12}}\right|$$

2.8 To verify the conditions for unilateralization first the complete circuit shown in Fig. 2.3 has to be analysed with respect to its resulting parameter $y_{\Sigma 12}$ (Appendix C).

The whole circuit can be described by the matrix equation

$$\boldsymbol{Y}_\Sigma = (\boldsymbol{z}_s)^{-1} + \boldsymbol{Y}_P$$

if

$$\boldsymbol{z}_S = \begin{bmatrix} z_{11} & z_{12} \\ z_{21} & z_{22} + jX_1 \end{bmatrix}$$

describes the active fourpole $\boldsymbol{Z} = (z_{kl})$ together with X_1 at the output, and

$$\boldsymbol{Y}_P = \begin{bmatrix} jB_2 & -jB_2 \\ -jB_1 & jB_2 \end{bmatrix}$$

represents X_2.

For neutralization the feedback parameter $y_{\Sigma 12}$ has to be zero, which consists of the sum of

$$Y_{S12} = \frac{-z_{S12}}{\Delta_{ZS}} \quad \text{and} \quad (-jB_2)$$

$$\Delta z_S = z_{S11} z_{S22} - z_{S12} z_{S21}$$

Therefore

$$0 = \frac{-z_{21}}{z_{11} z_{22} + jz_{11} X_1 - z_{12} z_{21}} - jB_2$$

is the condition for neutralization. Introducing the y-parameter of the active fourpole leads to

$$jB_2 = \frac{Y_{12}}{1 + jX_1 Y_{22}}$$

Solving for the real and imaginary part it follows that

$$\frac{1}{X_1} = B_1 = \frac{b_{12} g_{22}}{g_{12}} - b_{22}$$

and

$$\frac{1}{X_2} = B_2 = b_{12} - \frac{b_{22} g_{12}}{g_{22}}$$

($y_{kl} = g_{kl} + jb_{kl}$ (Appendix A)).

As can be concluded for any set of parameters (y_{kl}) a combination of X_1, X_2 can be found allowing lossless neutralization.

The corresponding unilateral gain

$$U = \frac{|y_{21} - y_{12}|^2}{4(g_{11} g_{22} - g_{12} g_{21})}$$

is the same for any choice of input and output of the same active device (common gate/base B; common source/emitter E; common drain/collector C). This can be concluded by transforming the parameters according to

$$y_{11\mathrm{E}} = y_{11\mathrm{B}} + y_{12\mathrm{B}} + y_{21\mathrm{B}} + y_{22\mathrm{B}}$$

$$y_{12\mathrm{E}} = -(y_{12\mathrm{B}} + y_{22\mathrm{B}})$$

$$y_{21\mathrm{E}} = -(y_{21\mathrm{B}} + y_{22\mathrm{B}})$$

$$y_{22\mathrm{E}} = Y_{22\mathrm{B}}$$

and

$$y_{11\mathrm{C}} = y_{11\mathrm{B}} + y_{12\mathrm{B}} + y_{21\mathrm{B}} + y_{22\mathrm{B}}$$

$$y_{12\mathrm{C}} = -(y_{11\mathrm{B}} + y_{21\mathrm{B}})$$

$$y_{21\mathrm{C}} = -(y_{11\mathrm{B}} + y_{12\mathrm{B}})$$

$$y_{22\mathrm{C}} = Y_{11\mathrm{B}}$$

The same equations are achieved, e.g. from

$$U = \frac{|y_{21\mathrm{E}} - y_{12\mathrm{E}}|^2}{4(g_{11\mathrm{E}} g_{22\mathrm{E}} - g_{12\mathrm{E}} g_{21\mathrm{E}})}$$

It follows that

$$U = \frac{|-(y_{21B} + y_{22B}) + (y_{12B} + y_{22B})|^2}{4\{(g_{11B} + g_{12B} + g_{21B} + g_{22B})g_{22B} - (g_{12B} + g_{22B})(g_{21B} + g_{22B})\}}$$

Rearranging leads to

$$U = \frac{|-y_{21B} + y_{12B}|^2}{4\{g_{11B}g_{22B} - g_{12B}g_{21B}\}}$$

CHAPTER 3

3.1 Corresponding to Fig. Ex3.1(a) the series connection of R_1, R_2 leads to

$$R_\Sigma = R = R_1 + R_2$$

The noise sources v_{n1}, v_{n2} are independent from each other, their correlation is zero. Therefore

$$\overline{(v_{n1} + v_{n2})^2} = \overline{v_{nl}^2} + \overline{v_{n2}^2}$$

and

$$v_{n\Sigma} = (v_{nl}^2 + v_{n2}^2)^{1/2}$$

It follows that

$$4kTRB = 4kT_1R_1B + 4kT_2R_2B$$

or

$$T_{eff} = T_1\frac{R_1}{R_1 + R_2} + T_2\frac{R_2}{R_1 + R_2}$$

As can be concluded, the effective noise temperature is dominated by the temperature of the largest resistance (same bandwidth B).

If the parallel connection is treated, the conductances have to be added,

$$G_\Sigma = G = G_1 + G_2$$

For the two independent noise sources

$$i_{nl} = (4kTG_1B)^{1/2}, \quad i_{n2} = (4kTG_2B)^{1/2}$$

it follows that

$$i_{n\Sigma} = (i_{n1}^2 + i_{n2}^2)^{1/2}$$

and T_{eff} becomes

$$T_{\text{eff}} = T_1 \frac{G_1}{G_1 + G_2} + T_2 \frac{G_2}{G_1 + G_2}$$

Therefore the resistor with the largest conductance dominates the effective temperature of the parallel connected resistors.

3.2 The following relationships can be extracted from the equivalent circuit after Strutt, where two noise current sources are implemented at input and output of the fourpole (Fig. Ex3.2(b)):

$$I_1 = y_{11}V_1 + y_{12}V_2 + i_1$$

$$I_2 = y_{21}V_1 + y_{22}V_2 + i_2$$

For the equivalent circuit with voltage and current source at the input (Fig. Ex.3.2(d)),

$$I_1 - i = y_{11}(V_1 - v) + y_{12}V_2$$
$$I_2 = y_{21}(V_1 - v) + y_{22}V_2$$

Therefore both configurations are equivalent as long as the relationships

$$i_1 = i - vy_{11} \quad \text{and} \quad i_2 = -vy_{21}$$

are valid.

To verify the equivalence with the configuration after Rothe and Dahlke (1956) (Fig. 3.2(e)), the noise current source has to be split into two parts; one noncorrelated part (i_n) and one correlated with the noise voltage v. This results in a correlation admittance Y_{cor} by setting

$$i = i_n + vY_{\text{cor}}$$

From Fig. Ex3.2(d),

$$V_1 = v + V_1'$$
$$I_1 = i + I_1' = vY_{\text{cor}} + i_n + I_1'$$

and from Fig. Ex3.2(e) it follows that

$$V_1 = v + V_1'$$

for the input current

$$I_1 = V_1 Y_{cor} + i_n + V'_1(-Y_{cor}) + I'_1$$
$$= (V_1 + V'_1)Y_{cor} + i_n + I'_1$$

Therefore

$$I_1 = vY_{cor} + i_n + I'_1$$

which proves the equivalence of the given circuits.

3.3 According to Fig. Ex3.3 the noise factor of two cascaded fourpoles is

$$F_\Sigma = \frac{P_{n2II}}{P_{n2IIO}} = \frac{P_{n2II}}{P_{nAVI}G_{TI}G_{effII}} = \frac{P_{n2II}}{kTBG_{TI}G_{effII}}$$

where P_{n2II} can be split into three parts,

$$P_{n2II} = P_{n2I}G_{effII} + P_{neII}$$
$$= kTBG_{TI}G_{effII} + P_{neI}G_{effII} + P_{neII}$$

where P_{neI} and P_{neII} are the additional excess noise powers of the first and second fourpole, respectively.

It follows that

$$F_\Sigma = 1 + \frac{P_{n2I}}{kTBG_{TI}} + \frac{P_{neII}}{kTBG_{TI}G_{effII}}$$

where the second term (first quotient) is the noise factor F_I of the first fourpole.

The third term can be written

$$\frac{P_{neII}}{kTBG_{TI}G_{effII}} = \frac{P_{neII}}{kTBG_{TII}} \frac{G_{TII}}{G_{TI}G_{effII}} = (F_{II} - 1)\frac{G_{TII}}{G_{TI}G_{effII}}$$

With the definition of the power gain

$$G_{TII} = \frac{P_{2II}}{P_{1AVII}}, \quad G_{TI} = \frac{P_{2I}}{P_{1AVI}}, \quad G_{effII} = \frac{P_{2II}}{P_{1II}}$$

the Friis formula follows with

$$F_\Sigma = F_I + (F_{II} - 1)\frac{P_{1AVI}}{P_{2AVI}} = F_I + (F_{II} - 1)\frac{1}{G_{AVI}}$$

It should be emphasized that the above formulation is only valid if the noise factor F_{II} is measured with a (thermal noisy) generator impedance equal to the output impedance of fourpole I. This is due to the definition of G_{TII} which is related to the power delivered from fourpole I under matched condition.

3.4 Applying the Friis formula, we have

$$F_{I,II} = F_I + \frac{F_{II} - 1}{G_{AVI}}$$

and

$$F_{II,I} = F_{II} + \frac{F_I - 1}{G_{AVII}}$$

if two noisy fourpoles are coupled together in the sequence 1, 2 or 2,1.
For $F_{I,II} < F_{II,I}$,

$$F_I - 1 + \frac{F_{II} - 1}{G_{AVI}} < F_{II} - 1 + \frac{F_I - 1}{G_{AVII}}$$

or

$$(F_I - 1)\left(1 - \frac{1}{G_{AVII}}\right) < (F_{II} - 1)\left(1 - \frac{1}{G_{AVI}}\right)$$

leading to

$$\frac{F_I - 1}{1 - (1/G_{AVI})} < \frac{F_{II} - 1}{1 - (1/G_{AVII})}$$

Therefore the noise measure

$$M = \frac{F}{1 - (1/G_{AV})}$$

is the relevant quantity if the sequence of amplifier stages is discussed regarding their overall noise performance: the stage with the lowest M-value has to be used as the front end, not that with the lowest F-value. The latter is only true if $G_{AV} \gg 1$.

3.5 The tube (Fig. Ex3.4, saturated thermionic diode) exhibits shot noise (section F.2) with

$$i_{\mathrm{nD}}^2 = 2qI{\cdot}B$$

The resistor $R_G(=1/G_G;\ R_G \gg R)$ exhibits thermal noise,

$$i_{\mathrm{nG}}^2 = 4\,kT_0G_GB$$

Both noise sources have to be connected in parallel (Exercise 3.1), leading to

$$i_{\mathrm{n}\Sigma}^2 = (2qI + 4kT_0G_G)B$$

This can be expressed as

$$i_{\mathrm{n}\Sigma}^2 = (F+1)i_{\mathrm{nG}}^2$$

with

$$F = \frac{2qI}{4kT_0G_G} = \frac{I}{2V_TG_G}$$

With $V_T \approx 25\,\mathrm{mV}$ at room temperature it follows that

$$F = \frac{I/A}{G_G/S}\frac{1}{0.05}$$

For $F = I(\mathrm{mA})$ it must be $G_G = 0.002\,\mathrm{S}$ for $R_G = 50\,\Omega$. This resistance would match exactly the transmission line with $Z_0 = 50\,\Omega$, without reflections.

Because the effective noise power is

$$i_{\mathrm{n}}^2 = F^* i_{\mathrm{nG}}^2$$

the effective noise temperature T_{eff} of the generator resistance R_G, being at $T = T_0$, is

$$T_{\mathrm{eff}} = F^* T_0$$

To indicate the effective noise factor in amperemetres, the corresponding values are $F^* = 1$ for $I = 0\,\mathrm{mA}$, $F^* = 2$ for $I = 1\,\mathrm{mA}$, etc.

With the given capacitance in parallel the RC time constant is

$$50\,\Omega{\cdot}10\,\mathrm{pF} = 0.5 \times 10^{-9}\,\mathrm{s},$$

leading to

$$2\pi f_{\mathrm{cutoff}} = 1/RC = 2 \times 10^9\,\mathrm{s}^{-1},\ \text{or}\ f_{\mathrm{cutoff}} \approx 300\,\mathrm{MHz}$$

CHAPTER 4

4.1 The circuit given in Fig. Ex4.1 uses voltage–voltage negative feedback. This results in the stabilization of the voltage gain G_V. With $V_4 = V_2$ the modified gain becomes

$$G'_V = \frac{V_2}{V'_1} = \frac{V_2}{V_3 + V_1} = \frac{V_2}{KV_2 + V_1} = \frac{G_V}{1 + KG_V}$$

With

$$\log(G'_V = \log(G_V) - \log(1 + KG_V)$$

it is

$$\frac{dG'_V}{G'_V} = \frac{dG_V}{G_V} - \frac{K\,dG_V}{1 + KG_V}$$

If a disturbance occurs resulting from an influencing quantity J, then the absolute dependence becomes

$$\frac{dG'_V}{dJ} = \frac{dG_V}{dJ}\left\{\frac{1}{1 + KG_V} - \frac{KG_V}{(1 + KG_V)^2}\right\} = \frac{dG_V}{G_V}\,\frac{1}{(1 + KG_V)^2}$$

The relative change which occurs is

$$\frac{dG'_V/G'_V}{dJ} = \frac{dG'_V}{dJ}\,\frac{1}{G'_V} = \frac{dG_V/G_V}{dJ}\,\frac{1}{1 + KG_V} < \frac{dG_V/G_V}{dJ}$$

To achieve a reduction of 10,

$$\frac{dG'_V/G'_V}{dG_V/G_V} = 0.1$$

so for $G_V = 100$

$$10 = 1 + KG_V = 1 + K\cdot 100 \quad \text{or} \quad K = 0.09$$

A simple circuit verification would be a transformer with the voltage ratio (Fig. 4.1(b))

$$\frac{V_3}{V_4} = \frac{9}{100}$$

4.2 The signal transfer from E to A can be described by

$$W = \frac{A}{E} = \frac{\sum a_\upsilon E^\upsilon}{E}$$

Therefore the feedback system exhibits

$$W' = \frac{W}{1 - rW} = \frac{\sum a_\upsilon E^{\upsilon-1}}{1 - r\sum a_\upsilon E^{\upsilon-1}}$$

Logarithmic differentiation leads to

$$\begin{aligned}\frac{\mathrm{d}W'}{W'} &= \frac{\sum a_\upsilon E^{\upsilon-2}}{\sum a_\upsilon E^{\upsilon-1}} + r\frac{\sum a_\upsilon E^{\upsilon-2}}{1 - r\sum a_\upsilon E^{\upsilon-1}} \\ &= \frac{\sum a_\upsilon E^{\upsilon-2}}{\sum a_\upsilon E^{\upsilon-1}} \frac{1}{1 - r\sum a_\upsilon E^{\upsilon-1}} \\ &= \frac{\mathrm{d}W}{W} \frac{1}{1 - rW}\end{aligned}$$

Higher order differential quotients lead to similar expressions representing the general reduction (for $r < 0$) or enhancement ($r > 0$) of the nonlinearities involved.

4.3 Each stage exhibits the gain (section 4.1.1)

$$G_0 = \frac{G_{00}}{1 + \mathrm{j}\Omega}$$

Therefore the three stages lead to the total gain

$$G = G_0^3 = \frac{G_{00}^3}{(1 + \mathrm{j}\Omega)^3}$$

If the feedback is arranged in one step over the three identical stages, then the modified gain becomes

$$G_\Sigma = \frac{G}{1 - rG} = \frac{G_{00}^3}{(1 + \mathrm{j}\Omega)^3 - rG_{00}^3}$$

If in contrast each of the three stages incorporates a feedback path, with the same feedback coefficient r, the modified gain is

$$G'_\Sigma = \left(\frac{G_0}{1 - rG_0}\right)^3 = \frac{G_{00}^3}{(1 - rG_{00} + \mathrm{j}\Omega)^3}$$

At low frequencies, $\Omega \ll 1$, the resulting gain is not equal to the first version, and at higher frequencies the frequency-dependent slope of the gain as well as the stability limit are considerably modified. This follows from the different denominators in the gain formula. Practical circuits often exhibit both feedback versions to realize sufficient feedback as well as broadband stability.

4.4 From

$$V_{GS} + RI_D = 0$$

(DC input circuit) it follows, for the resistance, according to Fig. Ex4.3(b) with the current $I_D = I_{DP} = 0.5\,\text{mA}$, that

$$R = \frac{-V_{GS}}{I_{DP}} = 200\,\Omega$$

and for the transconductance

$$g_m \approx \frac{\Delta I_D}{\Delta V_{GS}} = \frac{10\,\text{mA}}{1\,\text{V}} = 10\,\text{mS}$$

From the simplified circuit after Fig. Ex4.2(c),

$$V_E = \frac{1}{j\omega L + (1/j\omega C)} = -V_1 j\omega C$$

and

$$V_1 j\omega C - V_E\left(j\omega C + \frac{1}{R}\right) + g_m V_1 + I_E = 0$$

It follows, after some rearrangement, that

$$Y_E = G + g_m\left(1 - \frac{\omega C}{\omega C - (1/\omega L)}\right) + j\omega L\left(2 - \frac{\omega C}{\omega C - (1/\omega L)}\right)$$

or

$$Y_E = G + \frac{g_m}{1 - \omega^2 LC} + j\omega C\left(\frac{2 - \omega^2 LC}{1 - \omega^2 LC}\right)$$

For $I_m\{Y_E\} = 0$,

$$\omega_0^2 LC = 2$$

or

$$f_0 = \frac{1}{2\pi}\left(\frac{2}{LC}\right)^{1/2}$$

The corresponding real part becomes

$$\mathrm{Re}\{Y_E\} = G - g_m$$

or

$$\mathrm{Re}\{Y_E\} = 5\,\mathrm{mS} - 10\,\mathrm{mS} = -5\,\mathrm{mS}$$

To obtain the onset of oscillations, at the output terminals the sum of all conductances has to be smaller than zero,

$$\mathrm{Re}\{Y_E\} + G_L \leqslant 0$$

($G_L = 1/R_L$). Therefore the smallest resistance allowing for oscillations would be $R = 200\,\Omega$.

4.5 With the circuit configuration of Fig. Ex4.4 the $\boldsymbol{H}$ matrix of the active device is

$$\boldsymbol{H}_1 = \begin{bmatrix} h_{11} & h_{12} \\ h_{21} & h_{22} \end{bmatrix}$$

and the $\boldsymbol{H}$ matrix of the transformer (assumed to be ideal)

$$\boldsymbol{H}_2 = \begin{bmatrix} 0 & -x \\ +x & 0 \end{bmatrix}$$

This leads to

$$\boldsymbol{H}_\Sigma = \begin{bmatrix} h_{11} & h_{12} - x \\ h_{21} + x & h_{22} \end{bmatrix}$$

and the output admittance becomes for $Z_G = 0$ (short-circuit input)

$$Y_2 = h_{22} - \frac{(h_{12} - x)(h_{21} + x)}{h_{11}}$$

If a bipolar transistor is assumed with the simplified h-parameters $h_{12} = 0$ and $h_{21} = -1$, it follows that

$$Y_2 = h_{22} - \frac{x(1-x)}{h_{11}}$$

The optimum value is achieved for $x = 0.5$, leading to

$$Y_{2\text{opt}} = h_{22} - \frac{Y_{11}}{4}$$

($y_{11} = 1/h_{11}$). Because at normal bias conditions $|y_{11}| \gg 4|h_{22}|$ the negative admittance at the output becomes

$$Y_{\text{opt}} \approx -\frac{y_{11}}{4}$$

at the output terminals a negative conductance

$$G_{\text{N}} = -\frac{G_{\text{d}}}{4}$$

is generated as well as a negative capacitance

$$C_{\text{N}} = -\frac{C_{\text{d}}}{4}$$

($y_{11} = G_{\text{d}} + j\omega C_{\text{d}}$). With

$$G_{\text{d}} = \frac{I_{\text{E}}}{V_{\text{T}}}$$

(I_{E} is the DC emitter current, V_{T} the temperature voltage) the growth constant becomes

$$\sigma = \frac{|G_{\text{N}}| - G_{\text{L}}}{2C}$$

allowing for the generation of oscillations up to

$$G_{\text{Lmax}} = \frac{G_{\text{d}}}{4} = \frac{I_{\text{E}}}{4V_{\text{T}}}$$

With

$$C = C_{\text{L}} - \frac{C_{\text{d}}}{4}$$

the oscillation frequency is

$$f_0 = \frac{1}{2\pi(LC)^{1/2}}$$

With growing amplitude the negative admittance becomes reduced. This influences the oscillation frequency as well as the negative conductance. The final amplitude follows from the evaluation of the oscillatory characteristics (section D.2, Fig. D.3).

CHAPTER 5

5.1 Because of its high voltage amplification ($G_{V_o} \approx 10^4$) the ideal OP exhibits a negligible input voltage, $V_i \approx 0$ (Fig. Ex5.1). Therefore

$$V_B = -V_D = 5\,\text{V}$$

and the photocurrent becomes for the illumination of 500 lx

$$I_{ph} = -I_D = 5\,\mu\text{A}$$

With

$$-V_i + R_f I_{ph} + V_A = 0$$

the corresponding output voltage is $V_A = -5\,\text{V}$ ($R_f = 1\,\text{M}\Omega$).

The input impedance Z_e follows with

$$Z_e = \frac{V_e}{I_e} = \frac{V_i}{I_{ph}}$$

and

$$V_i = R_f I_{ph} + V_A$$

$$V_A = -G_V V_i$$

The result is

$$Z_e = \frac{R_f}{1 + G_V} \approx \frac{R_f}{G_V} = \frac{1\,\text{M}\Omega}{10^4} = 100\,\Omega$$

For the frequency limit (cutoff frequency f_c) the RC time constant of the combination $Z_e C_r$ is relevant. Therefore

$$Z_e 2\pi f_c C_r = 1$$

or

$$f_c = 15.9\,\text{MHz}$$

5.2 With

$$V_0 = R_V(I_B + I_Z) + V_Z$$

we have, for $I_Z \gg I_B$ (Fig. Ex5.2)

$$R_V = \frac{V_0 - V_Z}{I_Z} = 50\,\Omega$$

From Fig. Ex5.3(a) it follows for $I_L = I_E = 1\,\text{A}$, the voltage $V_{BE} = 0.7\,\text{V}$. The resulting voltage $V = V_Z - V_{BE}$ is 4.3 V. Thus the base current of $I_B = 9.9\,\text{mA}$ with

$$I_L = I_E = I_C + I_B$$

and

$$I_B = \frac{I_L}{1 + B}$$

The dynamic inner resistance becomes

$$R_i = -\frac{-\Delta V_{BE}}{\Delta I_E} = \frac{60\,\text{mV}}{0.6\,\text{A}} = 0.1\,\Omega$$

This value is strongly reduced if the operational amplifier (OP) is implemented. Again with $I_B = 9.9\,\text{mA}$,

$$I_B = I = g_m V_e$$

it follows that

$$V_e = \frac{I_B}{g_m} = \frac{9.9\,\text{mA}}{10^4\,\text{mS}} = 0.99\,\text{mV}$$

The voltage becomes

$$V = V_Z - V_e = 4.99901\,\text{V}$$

For $V_Z = \text{const.}$,

$$\Delta V = -\Delta V_e \quad \text{and} \quad \Delta V_e = \frac{\Delta I_B}{g_m} = \frac{\Delta I_L}{g_m(1 + B)}$$

Therefore the inner resistance is

$$R_i = \frac{1}{g_m(1+B)} = \frac{1\,V}{101 \times 10^4\,\text{mA}} = 0.99\,\text{m}\Omega$$

This is a reduction of a factor of 100.

5.3 The circuit with the CC device can be simplified for the analysis according to Fig. Ex5.4. Either the admittance function (*Y*, Fig. Ex5.4(b)) or the impedance function (*Z*, Fig. Ex.5.4(c)) can be used to derive the characteristic equation regarding stability or instability.

Using the *Z* version it is

$$z = -R_N + j\omega L + \frac{1}{j\omega C + (1/R_L)}$$

$$= \frac{R_L + (1 + j\omega R_L C)(j\omega L - R_N)}{1 + j\omega R_L C}$$

and therefore the characteristic equation is ($j\omega \rightarrow p$)

$$0 = R_L - R_N + p(L - R_N R_L C) + p^2 R_L C L$$

The roots are ($R_N = R_L$)

$$p_{1,2} = -\frac{L - R_N R_L C}{2 R_L C L} \pm \left(\frac{R_N - R_L}{R_L C L} + \left(\frac{L - R_N R_L C}{2 R_L C L} \right)^2 \right)^{1/2}$$

or ($R_N < R_L$)

$$p_{1,2} = -\frac{L - R_N R_L C}{2 R_L C L} \pm j \left(\frac{R_L - R_N}{R_L C L} - \left(\frac{L - R_N R_L C}{2 R_L C L} \right)^2 \right)^{1/2}$$

Therefore stability exists for

$$\sigma_{1,2} = -\frac{L - R_N R_L C}{2 R_L C L} < 0$$

or

$$R_L < \frac{L}{C R_N}$$

By setting the real and imaginary part of the complex impedance *Z* equal to zero, respectively, after some computation we have for the attenuation corner frequency

$$f_a = \frac{1}{2\pi R_L C}\left(\frac{R_L}{R_N} - 1\right)^{1/2}$$

and for the self resonant frequency

$$f_s = \frac{1}{2\pi R_L C}\left(1 - \frac{R_L^2 C}{L}\right)^{1/2}$$

The duality to the VC device follows from the correspondences

$$R_L \leftrightarrow G_L; \quad R_N \leftrightarrow G_N; \; L \leftrightarrow C$$

5.4 In the case of conventional amplifying devices the gain mechanism is based on variable ohmic components in the active device. However, under special circumstances varying reactances $L(t)$ and $C(t)$ also allow amplification. This parametric amplification is applicable for narrowband amplifiers at very high frequencies. Because reactances and not resistances are the active part of these amplifiers, they exhibit extremely low noise (Uenohara, 1960), especially if cooled (e.g. at 77 K).

The simplified circuit to be considered is shown in Fig. 5.29(b) (parallel resonant circuits at the signal frequency f_s, idler frequency f_i and pump frequency f_p). The variable capacitance is modulated by f_p (neglecting the weak additional influence of the signals at f_s or f_i):

$$C = C_0 + C_1 \cos \omega_p t$$

where C_0 is common to all resonant circuits and will not be considered further. The signals effective in the analysis are the voltages at f_s and f_i,

$$v = \hat{v}_s \cos(\omega_s t + \varphi_s) + \hat{v}_i \cos(\omega_i t + \varphi_i)$$

where the current through the variable part of C is

$$i = \dot{Q}_c = \frac{d}{dt}(Cv) = v\frac{dC}{dt} + C\frac{dv}{dt}$$

This leads to

$$\begin{aligned} i = & -C_1\omega_p \sin \omega_p t \hat{v}_s \cos(\omega_s t + \varphi_s) \\ & -C_1\omega_p \sin \omega_p t \hat{v}_i \cos(\omega_i t + \varphi_i) \\ & -C_1 \cos \omega_p t \omega_s \hat{v}_s \sin(\omega_s t + \varphi_s) \\ & -C_1 \cos \omega_p t \omega_i \hat{v}_i \sin(\omega_i t + \varphi_i) \end{aligned}$$

With

$$\sin\alpha\cos\beta = \tfrac{1}{2}\{\sin(\alpha+\beta) + \sin(\alpha-\beta)\}$$

it follows that

$$\begin{aligned}
-\frac{2i}{C_1} = {} & \omega_p \hat{v}_s \sin\{(\omega_p + \omega_s)t + \varphi_s\} \\
& + \omega_p \hat{v}_s \sin\{(\omega_p - \omega_s)t - \varphi_s\} \\
& + \omega_p \hat{v}_i \sin\{(\omega_p + \omega_i)t + \varphi_i\} \\
& + \omega_p \hat{v}_i \sin\{(\omega_p - \omega_i)t - \varphi_i\} \\
& + \omega_s \hat{v}_s \sin\{(\omega_p + \omega_s)t + \varphi_s\} \\
& + \omega_s \hat{v}_s \sin\{(-\omega_p + \omega_s)t + \varphi_s\} \\
& + \omega_i \hat{v}_i \sin\{(\omega_p + \omega_i)t + \varphi_i\} \\
& + \omega_i \hat{v}_i \sin\{(-\omega_p + \omega_i)t + \varphi_i\}
\end{aligned}$$

If the frequencies are chosen as

$$\omega_p = \omega_i + \omega_s$$

and therefore

$$\omega_p - \omega_i = \omega_s, \quad \omega_p - \omega_s = \omega_i$$

we have (neglecting the signals at $\omega_p + \omega_i$, $\omega_p + \omega_s$)

$$\begin{aligned}
-2\frac{i}{C_1} = {} & \omega_p \hat{v}_s \sin(\omega_i t - \varphi_s) \\
& + \omega_p \hat{v}_i \sin(\omega_s t - \varphi_i) \\
& - \omega_s \hat{v}_s \sin(\omega_i t - \varphi_s) \\
& - \omega_i \hat{v}_i \sin(\omega_s t - \varphi_i)
\end{aligned}$$

The relevant resulting currents at f_s and f_i are therefore given by

$$-\frac{2i}{C_1} = \omega_i \hat{v}_s \sin(\omega_i t - \varphi_s) + \omega_s \hat{v}_i \sin(\omega_s t - \varphi_i)$$

Here complex notation has to be introduced: with

$$i_{\mathrm{i}} = -\frac{C_1}{2}\omega_{\mathrm{i}}\hat{v}_{\mathrm{s}}\sin(\omega_{\mathrm{i}}t - \varphi_{\mathrm{s}})$$
$$= -\frac{C_1}{2}\omega_{\mathrm{i}}\hat{v}_{\mathrm{s}}\cos\left\{\omega_{\mathrm{i}}t - \frac{\pi}{2} - \varphi_{\mathrm{s}}\right\}$$

the corresponding current phasor at f_{i} becomes

$$I_{\mathrm{i}} = -\frac{\omega_{\mathrm{i}}C_1}{2}V_{\mathrm{s}}^*\,\mathrm{e}^{-\mathrm{j}\pi/2} = \mathrm{j}\omega_{\mathrm{i}}C_1\frac{V_{\mathrm{s}}^*}{2}$$

Because at this frequency no external generator exists, the total current must be zero,

$$I_{\mathrm{i}} + I_{\mathrm{i0}} = 0$$

With

$$I_{\mathrm{i0}} = Y_{\mathrm{i}}V_{\mathrm{i}}$$

we have

$$-Y_{\mathrm{i0}}V_{\mathrm{i}} = \mathrm{j}\omega_{\mathrm{i}}C_1\frac{V_{\mathrm{s}}^*}{2}$$

or

$$V_{\mathrm{i}} = V_{\mathrm{s}}^*\frac{-\mathrm{j}\omega_{\mathrm{i}}C_1}{2Y_{\mathrm{i0}}}$$

Similarly at f_{s},

$$i_{\mathrm{s}} = -\frac{C_1}{2}\omega_{\mathrm{s}}\hat{v}_{\mathrm{i}}\sin(\omega_{\mathrm{s}}t - \varphi_{\mathrm{i}})$$
$$= -\frac{C_1}{2}\omega_{\mathrm{s}}\hat{v}_{\mathrm{i}}\cos\left\{\omega_{\mathrm{s}}t - \frac{\pi}{2} - \varphi_{\mathrm{i}}\right\}$$

leading to

$$I_{\mathrm{s}} = -\frac{C_1}{2}V_{\mathrm{i}}^*\,\mathrm{e}^{-\mathrm{j}\pi/2} = \mathrm{j}\omega_{\mathrm{s}}C_1\frac{V_{\mathrm{i}}^*}{2}$$

With

$$V_{\mathrm{i}}^* = \left[\frac{-\mathrm{j}\omega_{\mathrm{i}}C_1V_{\mathrm{s}}^*}{2Y_{\mathrm{i0}}}\right]^*$$
$$= \mathrm{j}\frac{\omega_{\mathrm{i}}C_1V_{\mathrm{s}}}{2Y_{\mathrm{i0}}^*}$$

we have

$$I_s = \frac{j\omega_s C_1}{2} \frac{j\omega_i C_1}{2Y^*_{i0}} V_s$$

$$= -\frac{\omega_s \omega_i C_1^2}{4} \frac{1}{Y^*_{i0}} V_s = Y_{ns} V_s$$

Therefore a negative admittance is generated at the signal frequency which leads at resonance to

$$Y_{ns} = -G_N = \frac{-\omega_s \omega_i C_1^2}{4G_{i0}}$$

The RF powers involved are

$$P(f_i) = P_i \quad \text{and} \quad P(f_s) = P_s$$

both of which have to be supplied by the pump generator (Rauscher and Tucker, 1977) at $f = f_p$:

$$P_i = -\tfrac{1}{2} R_e\{I_i^* V_i\}$$

which is, with

$$I_i = j\omega_i C_1 \frac{V_s^*}{2} \quad \text{and} \quad V_i = \frac{-j\omega_i C_1 V_s^*}{2Y_{i0}}$$

given as

$$P_i = \frac{1}{2} R_e \left\{ \frac{\omega_i^2 C_1^2 |V_s|^2}{4Y_{i0}} \right\}$$

The power delivered at f_s becomes

$$P_s = \tfrac{1}{2} R_e\{V_s^* I_s\}$$

$$= -\tfrac{1}{2} R_e\{|V_s|^2 Y_{ns}\}$$

$$= \tfrac{1}{2}|V_s|^2 \,\mathrm{Re}\left\{ \frac{\omega_s \omega_i C_1^2}{4Y^*_{i0}} \right\}$$

$$= \tfrac{1}{2}|V_s|^2 \,\mathrm{Re}\left\{ \frac{\omega_s \omega_i C_1^2}{4Y_{i0}} \right\}$$

Therefore

$$\frac{P_s}{P_i} = \frac{\omega_s}{\omega_i}$$

The total power is

$$P(f_p) = P_p = P_s + P_i = P_s\left(1 + \frac{P_i}{P_s}\right) = P_s\left(1 + \frac{\omega_i}{\omega_s}\right) = P_s\frac{\omega_p}{\omega_s}$$

which leads to the Manley–Rowe relationship

$$\frac{P_p}{\omega_p} = \frac{P_s}{\omega_s} = \frac{P_i}{\omega_i}$$

The unexpected fact is that the accepted pump power P_p is dependent on the strength of the input signal. With

$$P_p = \frac{|V_s|^2}{2}\left[\mathrm{Re}\left\{\frac{\omega_i^2 C_1^2}{4Y_{i0}} + \frac{\omega_i\omega_s C_1^2}{4Y_{i0}}\right\}\right]$$

$$= |V_s|^2 \frac{\omega_p\omega_i C_1^2}{8}\mathrm{Re}\{Z_{i0}\}$$

for $v_s = 0$, no pump power is distributed in the circuit

5.5 With

$$Y_e = G_{de} + j\omega C_e, \quad Y_c = G_c + j\omega C_c, \quad Y_o = G_o + j\omega C_o$$

and the definition of the y-parameters

$$y_{11} = \left.\frac{I_1}{V_1}\right|_{V_2=0} \quad y_{12} = \left.\frac{I_1}{V_2}\right|_{V_1=0}$$

$$y_{21} = \left.\frac{I_2}{V_1}\right|_{V_2=0} \quad y_{22} = \left.\frac{I_2}{V_2}\right|_{V_1=0}$$

we have

$$y_{11} = Y_e + Y_c \quad y_{12} = -Y_c$$

$$y_{21} = g_m - Y_c \quad y_{22} = Y_o + Y_c$$

Because for a conventional bipolar transistor under normal bias conditions

$$|Y_e| \gg |Y_c| \quad g_m \gg |Y_c| \quad |Y_o| > |Y_e|$$

a simplified inner π equivalent circuit can be composed by

$$y_{11} \approx Y_e$$
$$y_{12} \approx -Y_c$$
$$y_{21} \approx g_m$$
$$y_{22} \approx Y_0$$

If in addition a base resistance R_b is taken into account, the y-parameters are modified to y'_{kl}

For a short-circuit at the input, in front of R_b, a residual voltage V results which influences via the transconductance the current balance (Fig. Ex5.5(c)). Furthermore the applied input voltage is only partly active regarding the voltage V. Therewith a reduction factor

$$\frac{1}{1 + R_b y_{11}}$$

is relevant, and the modified parameters become

$$y'_{11} = y_{11} \frac{1}{1 + R_b y_{11}}$$
$$y'_{12} = y_{12} \frac{1}{1 + R_b y_{11}}$$
$$y'_{21} = y_{21} \frac{1}{1 + R_b y_{11}}$$

In case of y'_{22} the formula becomes even more complicated because at short-circuit input the inner voltage V is not zero. It follows that

$$y'_{22} = y_{22} - R_b \frac{y_{12} y_{21}}{1 + R_b y_{11}}$$

5.6 In the first moment after applying the input pulse, the voltage across G_{de}, C_{de}, remains constant because of the capacitive load of the inner G_{de}/C_{de} combination (Figs Ex5.6, S5.1). Therefore a voltage step occurs at the outer base contact,

$$\Delta V_{BE}|_{t=0} = \Delta V_1 = \frac{R_b}{R_b + R_i} \Delta V_E$$

The same (but negative) amount occurs at turn-off of the input signal at

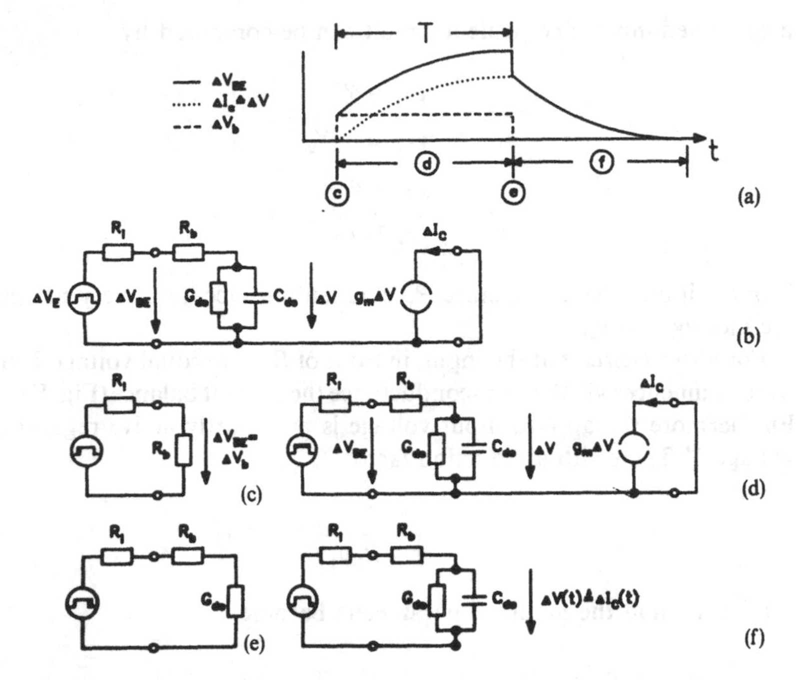

Fig. S5.1 Equivalent circuits for the different switching phases. (a) $\Delta V_{BE}(t)$ with indication of ΔI_C and ΔV_b; (b) equivalent circuit; (c) turn-on phase, $\Delta V = 0$; (d) on state (exponential growth); (e) turn-off, $\Delta V_E = 0$; (f) exponential decay.

$t = T$ (Fig. S5.1(a)). The measurable voltage step ΔV_1 depends on the value of R_b and is used for its determination.

The following slope of the transient function results from the charging of C_{de}, where the inner voltage ΔV grows exponentially. The characteristic time constant is given by $\tau = C_{de}R$, where R is the total resistance in parallel to C_{de}

$$R = \frac{R_b + R_i}{1 + (R_b + R_i)G_{de}}$$

The value of G_{de} is known because of the given bias condition,

$$G_{de} = \frac{I_E}{V_T}$$

where I_E is the emitter current and V_T the temperature voltage.

With the computed value of $\tau = t_d$, by inspection of $\Delta V_{EB}(t)$ on a curve tracer, it follows that

$$C_{dc} = \frac{\tau}{R}$$

5.7 Corresponding to Fig. Ex5.7 we have

$$i_{1E} = -i_{1B} - i_{2B}$$
$$v_{1E} = -v_{1B}$$
$$i_{2E} = i_{2B}$$
$$v_{2E} = v_{2B} - v_{1B}$$

This has to be introduced into the h-matrix of the common emitter connection

$$v_{1E} = h_{11E} i_{1E} + h_{12E} v_{2E}$$
$$i_{2E} = h_{21E} i_{1E} + h_{22E} v_{2E}$$

(The same procedure is valid for any other three-terminal device, e.g. an FET.) Thus it follows that

$$-v_{1B} = h_{11E}(-i_{1B} - i_{2B}) + h_{12E}(v_{2B} - v_{1B})$$
$$i_{2B} = h_{21E}(-i_{1B} - i_{2B}) + h_{22E}(v_{2B} - v_{1B})$$

This has to be brought into the form

$$v_{1B} = h_{11B} i_{1B} + h_{12B} v_{2B}$$
$$i_{2B} = h_{21B} i_{1B} + h_{22B} v_{2B}$$

The result after some computation is

$$h_{11B} = \frac{h_{11E}}{N_E}$$

$$h_{12B} = \frac{-h_{12E} + \Delta_{hE}}{N_E}$$

$$h_{21B} = -\frac{h_{21E} + \Delta_{hE}}{N_E}$$

$$h_{22B} = \frac{h_{22E}}{N_E}$$

where

$$N_E = 1 - h_{12E} + h_{21E} + \Delta_{hE}$$
$$\Delta_{hE} = h_{11E}h_{22E} - h_{12E}h_{21E}$$

5.8 In Fig. Ex5.8 different versions are shown. To evaluate the circuit EB/BB first the combination of the active device (e.g. a bipolar transistor) and the transformer has to be arranged as a parallel–series coupling of two fourpoles (Fig. Ex5.9). With the p-matrix of the active device

$$\boldsymbol{P}_I = \begin{bmatrix} p_{11} & p_{12} \\ p_{21} & p_{22} \end{bmatrix}$$

and that of the ideal transformer

$$\boldsymbol{P}_{II} = \begin{bmatrix} 0 & -x \\ -x & 0 \end{bmatrix}$$

the complete set of parameters follows by adding both matrices.

The result is

$$\boldsymbol{P}_\Sigma = \boldsymbol{P}_I + \boldsymbol{P}_{II} = \begin{bmatrix} p_{11} & p_{12} - x \\ p_{21} - x & p_{22} \end{bmatrix}$$

which shows the dependence on the connection point at the transformer, voltage ratio $1{:}x$. For $x = 0$ the common emitter connection is established, for $x = 1$ the common base connection, any intermediate connection becomes possible, allowing for changing the gain and the input and output resistance (the resulting quantities can be computed by applying the formulae of Appendix A).

If the transistor is given by its h parameters, $h_{11} = h_i, h_{12} = h_r$, $h_{21} = h_m$, $h_{22} = h_o$, the corresponding p-parameters are

$$p_{11} = \frac{h_{22}}{\Delta_h} \quad p_{12} = \frac{-h_{12}}{\Delta_h}$$
$$p_{21} = \frac{-h_{21}}{\Delta_h} \quad p_{22} = \frac{h_{11}}{\Delta_h}$$

where $\Delta_h = h_{11}h_{22} - h_{12}h_{21}$.

The back-transformation into the h-matrix of the combined fourpoles follows via

$$h_{11}=\frac{p_{22}}{\Delta_p} \quad h_{12}=\frac{-p_{12}}{\Delta_p}$$

$$h_{21}=\frac{-p_{21}}{\Delta_p} \quad h_{22}=\frac{p_{11}}{\Delta_p}$$

where the p-parameters are those of the combined circuit.

$$\Delta_p=\Delta_{p\Sigma}=p_{11\Sigma}p_{22\Sigma}-p_{12\Sigma}p_{21\Sigma}$$

Therefore

$$H=\begin{bmatrix} \dfrac{h_{11}}{\Delta_h\Delta_{p\Sigma}} & \dfrac{h_{12}+x\Delta_h}{\Delta_h\Delta_{p\Sigma}} \\ \dfrac{h_{21}+x\Delta_h}{\Delta_h\Delta_{p\Sigma}} & \dfrac{h_{22}}{\Delta_h\Delta_{p\Sigma}} \end{bmatrix}$$

with

$$\Delta_h\Delta_{h\Sigma}=1-x(h_{12}+h_{21})-x^2\Delta_h$$

For $x=0$ the fourpole parameters of the common emitter configuration (EC) occur, whereas for $x=1$ the h-parameters of the common base configuration occur (BB, with inverted input).

5.9 Because no further transfer function exists besides the current source at the output (Fig. Ex5.10), $S_{12}=0$.

The value of S_{11} is identical with the input reflection factor,

$$S_{11}=\frac{1-y_{11}Z_0}{1+y_{11}Z_0}$$

where Z_0 is the characteristic impedance of the transmission line.

Because of the infinitely high output impedance, $S_{22}=1$.

The wave parameter S_{21} is given by

$$S_{21}=\left.\frac{b_2}{a_1}\right|_{a_2=0}$$

With the relevant waves

$$a_1=\frac{1}{2}\left(\frac{V_1}{(Z_0)^{1/2}}+y_{11}V_1(Z_0)^{1/2}\right)$$

$$a_2 = \frac{1}{2}\left(\frac{V_2}{(Z_0)^{1/2}} - y_{21}V_1(Z_0)^{1/2}\right)$$

$$b_2 = \frac{1}{2}\left(\frac{V_2}{(Z_0)^{1/2}} + y_{21}V_1(Z_0)^{1/2}\right)$$

and $a_2 = 0$,

$$\frac{V_2}{(Z_0)^{1/2}} = y_{21}V_1(Z_0)^{1/2}$$

and therefore

$$S_{21} = \frac{2y_{21}V_1Z_0}{V_1 + y_{11}V_1Z_0} = \frac{2y_{21}Z_0}{1 + y_{11}Z_0}$$

5.10 By introducing the given parameters the gain of the amplifier shown in Fig. Ex5.11 can be computed. The relevant quantity is the quotient b_3/a_1, which follows from the inspection of the whole circuit including the circulator.

With its scattering matrix it is corresponding to Fig. Ex5.9

$$b_1 = \tau a_3$$
$$b_2 = \tau a_1$$
$$b_3 = \tau a_2$$

Because at port 2 it is

$$\frac{a_2}{b_2} = r$$

it follows

$$b_3 = \tau r b_2 = \tau^2 r a_1$$

Therefore the power gain is

$$G = \left|\frac{b_3}{a_1}\right|^2 = |\tau^2 r|^2$$

With

$$\frac{G}{\text{dB}} = 10 \log G = 20 \log|\tau^2 r|$$

it becomes

$$\frac{G}{\text{dB}} = 20\log(0.81\cdot 10)$$

Because $\log(8.1) = 0.908$,

$$G = 18.16\,\text{dB}$$

CHAPTER 6

6.1 Figure Ex6.2 gives the dependencies which have to be drawn first. The charge Q_S which has to be removed in the time interval $0 < t \leqslant t_S$ (storage time) is

$$Q_S = qA\frac{P_F}{2}x_S$$

where

$$\frac{x_S}{W - x_S} = \frac{P_F}{P_R}$$

Therefore with

$$I_F = qAD_p\frac{P_F}{W}, \quad I_R = qAD_p\frac{P_R}{W}$$

and

$$\frac{P_R}{P_F} = \frac{I_R}{I_F}$$

it follows that

$$Q_S = I_F\frac{W^2}{2D_p}\frac{1}{1 + (I_R/I_F)}.$$

This charge must be extracted by $i(t) \approx -I_R$ in the first time interval $0 < t \leqslant t_S$, where at $t = 0$ it is

$$i(t=0) = -I_{RO} = -\frac{|V_R| + V_F}{R_R}$$

and at $t = t_S$ exactly (Fig. E6.1(a))

$$i(t = t_S) = -I_R = -\frac{|V_R|}{R_R}$$

Therefore

$$I_R \approx \frac{Q_S}{t_S}$$

and the storage time becomes

$$t_S = \frac{W^2}{2D_p} \cdot \frac{1}{(I_R/I_F)(1 + (I_R/I_F))}$$

where the characteristic time constant is the diffusion time constant t_d equal to the transit time,

$$\frac{W^2}{2D_p} = t_d = t_t$$

In the second phase the (assumed) exponential decay takes place. The total charge which is stored in the base at $t = t_S$ is extracted partially via the junction (Q_D) and partially by recombination at the base contact (Q_C) at $x = W$.

With

$$i(t - t_S) = -I_R e^{-(t - t_S)/\tau} \quad (t \geqslant t_S)$$

and

$$Q_D = -\int_{t_S}^{\infty} i(t)\mathrm{d}t = I_R \tau$$

it follows that

$$\tau = t_t \frac{1}{1 + (I_R/I_F)}$$

because

$$Q_D = Q_R \frac{x_S}{W} = Q_R \frac{1}{1 + (I_R/I_F)} = t_t \frac{I_R}{1 + (I_R/I_F)}$$

Again the quotient I_R/I_F determines the behaviour as in the case of the storage phase.

6.2 For simplicity a short base p–n diode is assumed, with base width $W <$ diffusion length L_D of the in-diffusing minority carriers. Under this condition a triangular carrier distribution of the majority carriers move in via the base contact to achieve the afforded (quasi) charge neutrality.

For current turn-on the input current I_F is constant, and in any time interval Δt of the filling phase the same amount of minority carriers

$$\Delta Q = I_F \Delta t$$

flows into the base (Fig. Ex6.4(a)). Therefore the time needed to reach the final (always triangular) distribution with the final charge Q_0 must be

$$t_{onI} = \frac{Q_0}{I_F}$$

The total charge is

$$Q_0 = qAW\frac{P_F}{2}$$

(Exercise 6.1), and the current is

$$I_F = qAD_p\frac{P_F}{W}$$

Therefore the turn-on time is

$$t_{onI} = \frac{W^2}{2D_p} = t_t$$

which is equal to the transit time.

Figure Ex6.4(b) shows the time-dependent distribution for voltage turn-on. In the first moment the initial current is

$$I_{F0} \approx \frac{V_F}{R_b}$$

which is only limited by the base resistance R_b. Because throughout the loading time the input current is greater than the final steady state current I_F, the voltage turn-on time is considerably reduced with respect to current turn-on, $t_{onV} < t_{onI}$. If a charging current

$$i(t) = I_F + (I_{F0} - I_F)e^{-t/\tau}$$

is assumed with exponential decay, the total charge follows with

$$Q_0 = I_F t_t = \int_0^{t_{onV}} i\,\mathrm{d}t \approx I_F t_{onV} + \left(\frac{V_F}{R_b} - I_F\right)\tau$$

under the assumption that $t_{onV} \ll \tau$. With this equation t_{on} can be evaluated.

If a turn-on time of $t_{onV} \approx 3\tau$ is assumed, it follows that

$$\tau = t_t \frac{1}{2 + (V_F/I_F R_b)}$$

and therefore

$$t_{onV} = t_t \frac{3}{2 + (V_F/I_F R_b)} = t_t \frac{3}{2 + (I_{F0}/I_F)} < t_{onI}$$

If a drift field is incorporated in the base to enhance the speed of the minority carriers in the base, the total charge becomes reduced. Instead of Q_0 the reduced charge Q_d is needed for the same final current I_E (Fig. Ex6.4(c)), which enhances the switching behaviour considerably; an extremely short base acts similarly. The resulting drift current needs only a small (constant) carrier density in the base is contrast to the diffusion current with the needed density gradient. This leads to a further reduction of the switching time.

Majority carrier diodes (Schottky diodes, modulation-doped diodes, triangular barrier diodes, Camel diodes) exhibit no such effects because these minority carriers do not contribute to the current (or only to a negligible extent).

6.3 In Fig. Ex6.5 the situation is sketched, using an n–p–n transistor as the example. The saturation densities n_E, n_C represent the forward currents I_{Ef}, I_{Cf}, respectively. The corresponding fictional density dependencies are also shown (....).

We have

$$I_C = -A_f I_{Ef} + I_{Cf}$$

and

$$I_E = -A_i I_{Cf} + I_{Ef}$$

where A_f, and A_i are the DC current amplification in the forward and reverse direction, respectively.

Because

$$I_B = -(I_E + I_C)$$

we have

$$I_B = -\{I_{Ef}(1 - A_f) + I_{Cf}(1 - A_i)\}$$

With

$$I_C = I_{Csat}\left(\approx \frac{V_B}{R_L}\right)$$

it follows that for the forward current of the collector diode,

$$I_{Cf} = A_f I_{Ef} + I_{Csat}$$

Introducing

$$Q_{B+} = -qAWn_c$$

where A is device area, and

$$I_{Cf} = -gAD_n \frac{n_c}{W}$$

we have

$$Q_{B+} = 2t_d I_{Cf}$$

because

$$\frac{W^2}{2D_n} = t_d$$

Therefore

$$Q_{B+} = 2t_d(A_f I_{Ef} + I_{Csat})$$

where the emitter current is ($V_{BE} > V_{BEsat}$)

$$I_{Ef} = I_{E0}e^{V_{BE}/V_T}$$

The extra charge in case of (over)saturation must flow out of the base layer before a decay of the outer current can take place. Because in this phase the current is about constant, it is

$$Q_{B+} = -t_s I_{Csat}$$

and the storage time becomes

$$t_s = -\frac{Q_{B+}}{I_{Csat}} = 2t_d\left(A_f \frac{I_{Ef}}{I_{Csat}} + 1\right)$$

The equivalent charge of majority carriers (holes) has also to leave the base, which happens via the base contact. A reduced storage time can be achieved if a higher reverse voltage is applied, equivalent to the large signal switching behaviour of a p–n diode (Exercise 6.1). This possibility is indicated in Fig. E6.5(d) with $t_s \rightarrow t'_s$. The current I_{Ceq} is the fictional current for V_{BEeff} if no limitation takes place.

The saturation factor follows by comparing the relevant input voltages for reaching saturation and above. With

$$I_{Efsat} = I_{E0} e^{V_{BEsat}/V_T}$$

$$I_{Ef} = I_{E0} e^{V_{BE}/V_T} > I_{Efsat}$$

it is

$$\frac{I_{Ef}}{I_{Efsat}} = e^{(V_{BE} - V_{BEsat})/V_T}$$

6.4 Without special circuit arrangements the decay of emitted light (LED, laser diode) depends on the residual radiative recombination of the electrons and holes which were stored before by positive biasing. This leads to a long emission tail after turn-off. In contrast the light turn-on time can be very small (a few picoseconds) depending on the electronic switching device used.

A reduced decay time can be achieved if a negative bias is applied directly after the emission phase, corresponding to the turn-off behaviour of a diode by applying the negative voltage (Exercise 6.2). Then the residual stored charges can be swept out suddenly, leading to a reduced decay time of the residual emission. This results in shorter optical pulse width (Δt) and allows so far the application of higher pulse rates.

To achieve this, a simple possibility is the use of a transmission line in parallel as shown schematically in Fig. Ex6.6, where the travelling electric turn-on pulse becomes reflected at the end, reaching the diode with a negative sign. The length of the line has to be adjusted in such a way that the inverted pulse arrives at the end of the intended pulse length Δt. A DC short at the end of this line can be tolerated as long as the turn-on period of the light-emitting device remains sufficiently short.

With the wave propagation velocity c_{eff} given by

$$c_{eff} \approx \frac{c_0}{(\varepsilon_r)^{1/2}}$$

where c_0 is the free space light velocity and ε_r the relative dielectric constant of the carrier material, the length l of the line has to be

$$l = \frac{C_{eff} \Delta t}{2}$$

6.5 The transfer function will be

$$i = a_1 v + a_2 v^2 + a_3 v^3$$

Two input functions are to be applied,

$$v = v_1 \cos \omega_1 t + v_2 \cos \omega_2 t$$

The second signal will be a strong and modulated carrier, $|v_2| \gg |v_1|$, with

$$v_2 = v_{20}(1 + m_2 \cos \Omega t)$$

After applying the relevant trigonometric formula and ordering,

$$\begin{aligned}
i = {} & \frac{a_2}{2}\{v_1^2 + v_2^2\} \\
& + \cos \omega_1 t\{a_1 v_1 + \tfrac{3}{4} a_3 v_1^3 + \tfrac{3}{2} a_3 v_1 v_2^2\} \\
& + \cos \omega_2 t\{a_1 v_2 + \tfrac{3}{4} a_3 v_2^3 + \tfrac{3}{2} a_3 v_1^2 v_2\} \\
& + \cos 2\omega_1 t\left\{\frac{a_2}{2} v_1^2\right\} + \cos 2\omega_2 t\left\{\frac{a_2}{2} v_2^2\right\} \\
& + \cos 3\omega_1 t\left\{\frac{a_3}{4} v_1^3\right\} + \cos 3\omega_2 t\left\{\frac{a_3}{4} v_2^3\right\} \\
& + \cos(\omega_1 - \omega_2)t\{a_2 v_1 v_2\} + \cos(\omega_1 + \omega_2)t\{a_2 v_1 v_2\} \\
& + \{v_1 \cos(2\omega_1 - \omega_2)t + v_1 \cos(2\omega_1 + \omega_2)t \\
& + v_2 \cos(\omega_1 - 2\omega_2)t + v_2 \cos(\omega_1 + 2\omega_2)t\}\tfrac{3}{4} a_3 v_1 v_2
\end{aligned}$$

Therefore at $\omega = \omega_1$ a component

$$v_1 \cos \omega_1 t\{\tfrac{3}{2} a_3 v_2^2\}$$

occurs, which contains the modulation of the signal at ω_2.
With

$$v_2^2 = v_{20}^2(1 + 2m_2 \cos \Omega t + m_2^2 \cos^2 \Omega t)$$

and

$$\cos^2 \Omega t = \frac{1 + \cos 2\Omega t}{2}$$

it follows that

$$\begin{aligned} i(\omega_1) &= i_{10}\cos\omega_1 t(1 + m_1\cos\Omega t + m_1'\cos 2\Omega t) \\ &= \cos\omega_1 t\{a_1 v_1 + \tfrac{3}{4}a_3 v_1^3 + \tfrac{3}{2}a_3 v_1 v_{20}^2(1 + \tfrac{1}{2}m_2^2) \\ &\quad + \tfrac{3}{2}a_3 v_1 v_{20}^2(2m_2\cos\Omega t + \tfrac{1}{2}m_2^2\cos 2\Omega t)\} \end{aligned}$$

The modulation transfer becomes

$$\frac{m_1}{m_2} = \frac{3a_3 v_{20}^2}{a_1 + \frac{3}{4}a_3\{v_1^2 + v_{20}^2(2 + m_2^2)\}} \approx \frac{3a_3 v_{20}^2}{a_1}$$

resulting in a modulation component of the former unmodulated (weak) signal $v_1 \cos\omega_1 t$.

With

$$i(\omega_1) = v_{10}\cos\omega_1 t(1 + m_1\cos\Omega t + m_1'\cos 2\Omega t)$$

we have

$$m_1 \approx \frac{3a_3 v_{20}^2}{a_1} m_2$$

Because

$$\frac{m_1'}{m_2} = \frac{\frac{3}{4}a_3 v_{20}^2 m_2}{a_1 + \frac{3}{4}a_3\{v_1^2 + v_{20}^2(2 + m_2^2)\}} \approx \frac{3a_3 v_{20}^2 m_2}{4a_1}$$

an additional distortion is involved, leading to a modulation with 2Ω,

$$m_1' = m_1 \frac{m_2}{4}$$

To determine the cross-modulation of a real device the quotient a_3/a_1 has to be measured. However, it should be mentioned that these measurements should not include measurements at harmonics of the base signals. The reason is that the small resulting signals can be distorted by nonlinearities of the generator signals used. A better advice is to measure the required coefficients at separated, nonharmonic frequencies; e.g. to determine the coefficient a_2 the output signal at $\omega_1 \pm \omega_2$ should be measured, and for a_3 one should use $2\omega_1 \pm \omega_2$ or $\omega_1 \pm 2\omega_2$.

Index

Page numbers appearing in **bold** refer to figures and page numbers appearing in *italic* refer to tables.